TECHNOLOGY-RICH TEACHING

Classrooms in the 21st Century

Gary L. Ackerman

University Press of America,® Inc.
Lanham · Boulder · New York · Toronto · Plymouth, UK

University Press of America,® Inc.
4501 Forbes Boulevard
Suite 200
Lanham, Maryland 20706
UPA Acquisitions Department (301) 459-3366

Unit A, Whitacre Mews, 26-34 Stannary Street,
London SE11 4AB, United Kingdom

Printed in the United States of America
British Library Cataloging in Publication Information Available

Library of Congress Control Number: 2015939459
ISBN: 978-0-7618-6608-4 (clothbound : alk. paper)
eISBN: 978-0-7618-6609-1

™ The paper used in this publication meets the minimum requirements of American National Standard for Information Sciences—Permanence of Paper for Printed Library Materials, ANSI Z39.48-1992

With thanks to the technicians, nurses, and doctors who helped my family and me that May morning in 2008.

Contents

Preface

Computers arrived in K-12 schools just as I was leaving them as a student. When I began my career as an educator five years later, the school in which I worked had multiple computer rooms, and they were replacing the computers they had originally purchased with new models. Coincident with my entering the profession was the emergence of the learning sciences, which began to (and continues to) elucidate the details of effective classrooms.

What educators knew about society and education to prepare students for that society along with what they knew about teaching and learning when I started my career is no longer true. We have seen computers and network technologies revise the skills and knowledge needed to be educated and the learning landscape has been reshaped.

The purpose of this book is to present a new vision of technology-rich teaching and learning. Several lenses for interpreting and frameworks for designing technology-rich teaching and learning are detailed. These ideas have supported my thinking and work with students and educators.

My career has been coincident with the education system's struggle to find a role for emerging technology and to renegotiate a role for itself in a society that is actively influenced by that technology. Because of technology and its influences on society and because of the new knowledge we have about teaching and learning, the schools from which I graduated, and in which I worked for much of my career, are now obsolete.

Introduction

MY RATIONALE

As a student, I attended a high school that had four computers available for students (my classmates' recollections confirm my memories). I was thoroughly unimpressed with the devices. I had fun playing the game in which I tried to hit my opponent's castle with projectiles. Ostensibly, the game was played to reinforce the lessons taught in math class. It didn't. I found little connection between what I was to study in college and computers.

Early in my studies to be a biology teacher, I found computers occasionally useful and halfway through my undergraduate studies it became obvious that computers were in my future as a teacher. As a result, I bought a computer (an Apple IIc) and enrolled in the "Computers in the Classroom" course offered in the college of education I attended. I used the Apple computer until I graduated and as I started my career teaching middle school science. Several years later, I came to a harsh realization. While I had been filling up big floppy disks with word processing and spreadsheet files and playing chess and occasionally running simulations for my students, a computer revolution had occurred, and I didn't even know what I didn't know about computers.

After I found a job teaching high school science, a physics-teacher colleague and I spent a summer revising our lab curricula so all of the data in our laboratory exercises could be collected using probes connected to IBM personal computers. Along with our students, we were able to focus on the trends in data as they were graphed in real-time; previously our laboratory activity had focused on the minutia of gathering data so trends did not focus our attention until later, often too late. Convinced computers could transform every teacher's classroom as it did our high school science classrooms, my colleague and I co-chaired the committee that wrote the school's first technology integration plan.

When interviewing for a position teaching middle school math at a different school, I volunteered to chair the committee the principal was convening to develop a plan to replace the computers in his building (which were comparable to the computers I had first used as an undergraduate student more than 10 years earlier). He offered me the job and the chair of the committee. I have been an educational technologist ever since. I have held several jobs in the intervening years, and in each I have assumed some level of responsibility for teaching with computers; coaching teachers on the use of technology in classrooms; and planning, installing, and maintaining computers and networks in schools.

I also became a graduate student. As a master's degree student at the local state college, I studied under the tutelage of professors who introduced me to qualitative research. I gathered data about my colleagues' experiences with technology and I studied the technology planning decisions made by school leaders in the counties around rural Vermont where I live. For doctoral studies, online learning was the only option as my family relied on my paycheck and the nearest in-person program was a three-hour drive away. I completed my doctoral coursework and research with a widely dispersed and diverse faculty, and all of our interaction occurred over the Internet (except for several telephone conversations with my committee chairperson).

As a distance learner, I did miss the opportunity to sit face-to-face with fellow graduate students, but I found my own cohort of technology-using educators in the New England League of Middle Schools. Until the demands of finishing my research caused me to resign, my time as an online graduate student coincided with my time serving and leading the NELMS Technology Committee, a dynamic group of professionals with whom I explored and shared and learned about the challenges of using computers in schools.

For my dissertation research, I sought to understand how leaders and technologists in fields other than K-12 education had responded to the emergence of computers. I found that some educators appeared to have independently discovered what was known by professionals in other fields and that some educators had begun to transform their teaching in ways that reflected what was happening in other fields. I discovered also, those educators appeared to be working in systems that were inhibiting rather than sustaining their efforts to transform education with technology. I am motivated to understand why systems inhibit rather than sustain those efforts and to remove any obstacles to those efforts.

I came to computers relatively late in life. It was not until I realized that computers would serve me and my students well that I began an active computing life; since then information and computer technology has found its way deep into my personal and my professional life. Still, I confess, a real disdain for computers. They can be unreliable and break when most-needed, and troubleshooting computers is a real hassle. I can usually resolve technical issues quickly and I design well-functioning systems, but it requires one attend to details that I find mundane. Still, I use technology for more and more. Still, also, I work regularly with adults and youngsters so they become competent and confident users of technology as we apply computers and information technology to understand and solve authentic problems.

My experiences have convinced me that computer-mediated communication is fundamental to life in the 21st century; humans adopt (with increasing rapidity) the information technologies in their environment and humans adapt their communication habits to the tools. Humans also exapt technology; they find new and unintended uses for technologies. In biology exaptations are those structures and functions that evolved for one purpose, but then were applied to a different purpose. The typical example is feathers, which were originally structures for thermoregulation, and later were adapted for flight. An analogous process occurs with technology; it is used for purposes unimagined (and unimaginable) by the inventors and designers.

The rapid adoption of, adaptation to, and exaptations of 21st century information technology that my research suggests has occurred in many fields has not occurred in education, however. Educators' reluctance to embrace emerging information technologies in a systematic manner can be blamed several factors including the precautionary principle—we are slow to accept any change until we are sure it is "the best for our students"—and our unwillingness to abandon familiar and safe practices that are deeply embedded in our existing culture. We have spent decades preparing with excessive precaution for a transformation that our students make in days. Our delays have also been caused by distractions arising from politicians, philanthropists, and business leaders seeking political advantage and profit from "educational reform." We are slow to adopt any changes not aligned with misguided direction from above.

Some may counter that school and technology leaders have provided what is necessary; computers are available, classrooms are connected, curriculum standards are in place, and teacher training is available. While these are all accurate observations, there is evidence the teaching and learning experienced by students is the same as it was prior to the arrival of the technology, and it would not change if the technology disappeared.

Children who were born in 1990 are now young adults and have always lived in a world with computers and the Internet. Both became ubiquitous as those children entered school in the middle of the 1990's. They have graduated from our schools and their children will be entering our schools in the coming years. We are no longer building classrooms for the first generation of digital natives, but for the second generation. Many conclude however that we still have schools that would be familiar to the parents of the children born in 1990. (This paragraph is largely autobiographical; my wife and I graduated from high school in 1983 and our first son was born in 1990. His brother was born in 1994. I found my children's classrooms very similar to the ones I attended; too similar.)

From the perspective of one who has spent his entire adult life teaching and learning with and about computers and information technology, I write with the purpose of supporting educators as they prepare for a new reality in education. We live and work in a new world and what educators did with previous generations of students can no longer be recognized as preparing students for their future. I have become convinced that classrooms in which students do not gain experience consuming and creating digital information and interacting over digital networks should be preserved and experienced as museum artifacts.

OVERVIEW OF THE BOOK

This book is about education that does not exist yet—at least in a systemic way. As a human endeavor, education has a history and its future is contingent upon that history. As the current generation of educators seeks to redefine the experience for digital generations, they will be overcoming inertia and precautions grounded in that history. In chapter one "How We Got Here," I recount the recent history of education related to teaching and technology. This is the version of education that I believe should be relegated to the museum.

Schools are organizations in which consensus is very difficult to achieve; the probability of sustaining commitment to any decision reached through consensus is lower still. Those decisions tend to be overturned when leadership changes or in response to other political influences. This is a source of real and reasonable frustration for teachers; it is unavoidable. In chapter two "Education as Wicked Technology," I describe unfamiliar lenses for understanding education. Through these lenses, the nature of school planning is described in a manner that explains and predicts planning decisions and the frustration that arises from those decisions.

Both society and technology exert strong and active influences on how human brains develop. The humans who enter classrooms as students have, even in their short lives, experienced culture and technology that is much different than that experienced by their parents and teachers. The nature of human brains and the factors that affect how they change, as well as the implications of new knowledge of brains, is the focus of chapter three "Learners."

In classrooms based on instructionism and designed to help students develop the academic skills necessary for industrial and information age society, learning tasks focused on transferring information from experts (primarily teachers and textbook authors) to students. Grounded in a more sophisticated understanding of learners and the society in which they will function, "Learner Tasks" for digital generations must use technology, focus on sharpening uniquely human skills, and be relevant for students. The nature of these tasks and some strategies for designing meaningful teaching around them are detailed in chapter four.

Schools are technology-rich places; students and teachers have access to computers and connections to the Internet are available. In chapter five, "Understanding the Technology Infrastructure in Schools" attention is paid to strategies for maintaining and sustaining the devices and systems students and teachers need.

Just as students must be prepared for new social realities, educators must constantly renew their work. In "Developing Educator Capacity," the unified theory of acceptance and use of technology is used to describe the types of experiences that prepare educators for their future. In addition, strategies for providing educators with a range of professional learning opportunities are presented.

The final chapter, "Developing System Capacity," turns attention to personal, leadership, and other strategies for ensuring teachers are supported in their work. Nature abhors a vacuum, and educators abhor reinventing the wheel. My colleagues are always seeking a solution designed and implemented by others to solve the same problem they face. While this is understandable, it rarely produces expected success. For this reason, this chapter (like the others) describes guidelines.

Those who seek recipes for creating a technology-rich classroom will be disappointed in this book. Those interested in the ideas, approaches, and strategies that have been useful to my colleagues and me over the last decades years will find the contents helpful.

Chapter 1
How We Got Here

In describing education as a social invention, Jerome Bruner observed, "each generation must define afresh the nature, direction, and aims of education to assure [that] freedom and rationality can be attained for a future generation" (1966, 22). He went on to detail how new discoveries in human growth and development lead to advances in learning theory and pedagogy and how changes in society contribute to changing expectations of educators. While Bruner's decades-old observation and his reasoning continue to describe factors necessitating the refreshing of education in the 21st century, the current state of education is far more complicated than Bruner could predict.

Late in the 20th century, just after Bruner's quotation was written, the field of cognitive science emerged and it continues to elucidate the details of human brain functions to an amazing level of detail. Learning science, which applies the discoveries of cognitive science to classrooms and other learning environments, is challenging much of the knowledge that was used to organize and inform 20th century schools.

When Bruner wrote in 1966, print was the dominant information technology just as it had been for centuries; electronic digital computers had yet to arrive as an option for most educators. They were still very large mainframe machines that required great expertise to operate. About a decade after Bruner wrote, computers entered the consumer market, and three decades after Bruner wrote, the Internet was ubiquitous. As connected computers moved from rooms to our desktops and now into our pockets, they brought access to effectively infinite digital information along. Now, the dominance of print is being challenged. Because networked computer technology is emerging as the dominant venue for information and interaction; fully participating in society requires expertise beyond the paper-based reading and writing and solving problems with mathematics that focused schooling in previous generations.

Like all human creations, education has a history and the future of education is contingent on that history. It is the source of institutional inertia and it limits (albeit temporarily) the potential futures of education. In this chapter, the pedagogy that characterized the industrial and information ages is reviewed, as is the history of computers in schools. The founding assumption of this book is that the version of technology-rich education that is common in the early 21st century is no longer sufficient to educate the members of the digital generations.

INSTRUCTIONISM: AN OBSOLETE PARADIGM

The experience and work we know as education exists within a paradigm, which is multidimensional and defines what we understand education to be and how it is to be accomplished. The paradigm is largely expressed through the practices of teachers and leaders; underlying those practices, however, is a collection of assumptions and theories about what humans learn, how they learn, and how to best teach them.

Educational researchers collect data in a systematic manner to refine the theories that predict and explain what we observe within the paradigm. As we will see, the paradigm of instructionism is increasingly challenged by findings in the cognitive and learning science research. Politicians and philanthropists, however, continue to make policy and funding decisions based on instructionism; this is an untenable situation.

Challenges to Instructionism

The dominant educational paradigm in the 20th century was based on easily recognizable ideas about how the human brain works and how to design classrooms to help human brains learn. R. Keith Sawyer (2006), a scholar from the University of Washington, articulated five assumptions in which curriculum and instruction has been grounded:

- curriculum comprises well-defined information and skills that represent necessary human knowledge
- the purpose of schools is to ensure students get the information and skills into their brains, thus become educated
- educators know how to deliver instruction so the curriculum is transferred into students' brains
- the most efficient instruction occurs from simple to complex
- the success of instruction can be measured with a test

Those assumptions appear to have been consistent with contemporary knowledge and experience: Literacy and numeracy in print-dominated societies were relatively slow to change and electronic media were largely the domain of popular culture, and most individuals consumed much more information than they created. Psychologists perceived the mind to be a container, and knowledge to be a cognitive phenomenon arising within an individual's brain. Measuring intelligence through tests had been underway since the early decades of the century.

These ideas were informed by behaviorist psychology that can be summarized by the adage "if one acts as if he or she knows 'something' than he or she knows it." These assumptions gave rise to the pedagogy known as direct instruction. Information is presented (commonly in a fast-paced manner) and students are expected to respond to questions in a manner consistent with the teacher's presentation; errors in students' response are corrected immediately (Burton, Moore, and Magliarno 2004).

Sawyer pointed out, however, that none of the assumptions about teaching and learning that underlie instructionism are supported by scientific evidence. Indeed, the discoveries of cognitive science and learning science contradict those assumptions. After challenging the assumptions of instructionism, Sawyer proposes deep learning (National Research Council 2000) as an alternative to the version of curriculum that supports instructionism. Among the assumptions in which deep learning are grounded are:

- appropriate curriculum depends on individual's existing knowledge as well as social context
- schools give students experiences within which they develop and refine skills for on-going learning
- through reflection, learners understand themselves as learners
- contextually complex problems of increasing relevance to students are the appropriate foci of curriculum
- learning is demonstrated through increased fluidity and flexibility as learners apply their expertise with greater ease and in more situations

Deeper learning posits a more active role for the learner in his or her learning. What the learner knows is the basis for future learning and learners summarize understandings through metaphors and other generalizations. The context in which learning occurs is also recognized as an essential aspect of deeper learning. Context includes both the social environment and the metacognitive focus.

From the perspective of the second decade of the 21st century, instructionism appears to be obsolete. It is based on inaccurate psychology and it does not provide necessary experiences. The basic tenants of instructionism are contradicted by cognitive science and alternatives are supported; we can better explain and predict what we observe using cultural-historical psychology rather than behaviorist psychology. Instructionism no longer produces the skill and knowledge necessary for the information and communication realities of society. In this book, instructionism is understood to be a marginalized pedagogy. There are some limited situations in which instructionism, including technology-mediated instructionism, is appropriate; but most of a 21st century student's experience in school should be designed to accomplish more sophisticated educational goals than can be accomplished through instruction.

Challenges from the 21st Century

The *Common Core State Standards* (CCSS) are the latest incarnation of the "things" students must known and be able to do once they attain a certain level of education. Ostensibly, these reflect the skills students will need to fully participate in society; the phrase "career and college readiness" is used throughout the CCSS materials in an attempt to communicate the comprehensive nature of the standards. Advocates claim students who score well on tests aligned with CCSS will be prepared for the future. There is little evidence this is true, and many educators and others have criticized the CCSS as a homogenizing factor (see for example Baker 2014, Borg 2014, Strauss 2014). Further, standards such as CCSS tend to be static, and are updated in a cycle that takes years to complete, which is contrary to the rapid evolution and development of information technology that we observe in society.

The focus of measuring educational outcomes with standardized tests can be traced to the efforts to measure generalized intelligence, *g*, using tests in the early 1900's. Not only is the concept of a unified measure of general intelligence contrary to the concept of intelligence as a multi-dimensional construct, but the validity, reliability, and predictive validity of standardized tests are dubious (Gould 1996, Sachs 1999).

For many educators, standards such as CCSS do not reflect the unique nature of their situation. A principal once observed, "The standards writers have amazing ego to think they know what my community needs better than my faculty and I know it." This principal argued that college and career readiness is much more sophisticated than it is constructed in the CCSS. He said, "The laundry list of items found in the standards does not reflect the complexities of what students need… at least according to what they thank and blame us for when they visit after graduating." He further commented on the obsolescence of the standards in technology, "We talked about them with our students, they laughed and thought we were talking about what elementary school students should do."

In a series of articles begun in 2007, Paul Beynon-Davies (2007, 2009a, 2009b, 2009c), a scholar from Cardiff University in the United Kingdom, define informatics to describe the social system, language, and technology through which all communication proceeds (see figure 1.1). The principal cited above is among the professionals who recognize the changing nature of both the social system and technology systems of 21st century informatics. Because of rapidly changing devices and information sources (along with vastly more information than can be used), global interaction that occurs nearly instantaneously, and cognition that can be downloaded to devices; the skills, knowledge, and habits that comprise the social system of the digital generations are unfamiliar to previous generations.

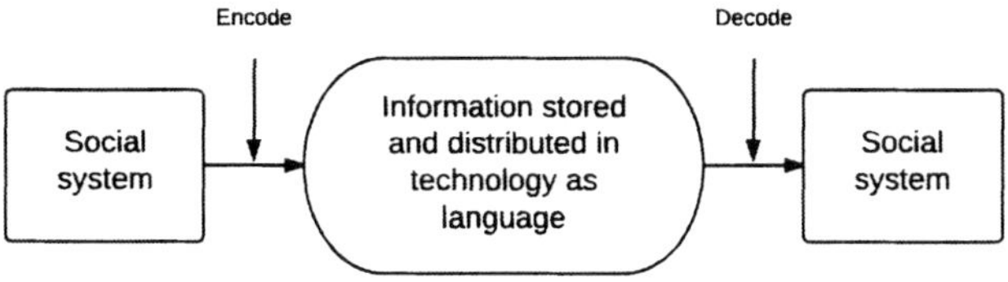

Figure 1.1. Informatics comprises social system and language and technology

Much of this book focuses on teaching and learning that is will replace instructionism. At this point, it is necessary to summarize the emerging technology-rich social system for which educators are preparing students.

Unpredictable Society

In his popular 2005 book *A Whole New Mind*, Daniel Pink concluded that the skills and types of thinking necessary for 20th century industrial and informational workers will no longer be sufficient for success in a global economy of abundance in which many of the tasks previously done by humans are now done by technology. Pink suggested 21st century skills and abilities are difficult to develop through instruction and are difficult to measure on tests. In the skills necessary for the future, there appears to be a rediscovery of those fundamental to human nature. Pink observed, "Back on the savannah, our cave-person ancestors weren't taking SAT's or plugging numbers into spreadsheets. But they were telling stories, building empathy, and designing innovations" (2005, 67). In this, Pink was suggesting that there would be an increasing differentiation between the tasks performed by technology and those performed by humans; those unique to humans being more important and relevant to solve society's problems and more central to education.

The changes suggested by Pink have already been observed in organizations. Jon-Arild Johannessen (2008), a scholar from the Norwegian School of Management, suggested that organizations gain competitive advantage in the global economy by developing innovative solutions to complex problems. Because of the pace at which technology evolves and the central role it plays in 21st century organizations, Johannessen concluded that sustained advantage is gained only through perpetual innovation. Innovative thinking can be neither imposed nor mandated and there is no recipe that can be followed by an organization's leaders to ensure that innovative solutions are produced. In innovative organizations, Johannessen concluded,

> The workforce will shift away from employees who have traditional, practical training backgrounds and towards an ever-increasing number of employees who have had a higher education and are theoretically well equipped. Such workers will be capable of working in a problem definition and problem-oriented manner and possess skills for both analysis and synthesis (407).

Olumuyia Asaolu (2002), a scholar in industrial and information engineering at the University of Tennessee, described several characteristics of interaction between individuals and within organizations that have been associated with the technologies they use; these characteristics allowed Asaolu to differentiate those individuals and organizations that were influenced by Fordist (Old) technologies from those influenced by ICT (New) technologies (see table 1.1). ICT is a commonly used abbreviation for information and computer technology. In general, the individuals and organizations that interact in a manner reflecting ICT (New) technologies are more flexible and have more dynamic interactions, whereas individuals who interact in a manner influenced by Fordist (Old) technologies are characterized by specialized and predictable interactions. Asaolu concluded as it becomes more ubiquitous and even more deeply embedded in everyday life, information and computer technology will exert even more powerful influences and increase the expectation that organizations adopt structures that reflect those influences.

Table 1.1. Characteristics of Fordist (old) and ICT (new) organizations

Fordist (old)	ICT (new)
Energy-intensive processes	Information-intensive processes
Standardized activities	Customized activities
Stable product mix	Flexible product systems
Dedicated facilities	Flexible production systems
Automation	Systemation
Single firms	Networks
Hierarchical management	Flat management
Specialized departments	Integrated
Product with service	Service with products
Centralized organization	Distributed organization
Specialized skills	Multi-skilling
Minimal training	Continuous retraining
Adversarial relationships between management and labor	Participative relationships between management and labor
Long-term full time employment	Flexible employment

adapted from Asaolu (2002)

In his 1996 book *The Rise of the Network Society*, Manuel Castells, a sociology professor at the University of California Berkeley who has held positions around the world, observed the late 20th century was marked by drastic changes in patterns of commerce and government that challenged long-established social norms around the world. Because of these technology-driven changes, new trends have been observed in a wide range of organizations;

Castells noted, "Technological innovation, and organizational change, focusing on flexibility and adaptability [are] absolutely critical in ensuring the speed and efficiency of the [organizational] restructuring" (1996, 19). Yochai Benkler, a professor at Yale Law School, observed that access to communication networks is associated with greater transparency in all aspects of culture and that transparency contributed to individuals having the power to participate in increasingly democratic governance. Benkler observed, "the change brought about by the networked information environment is deep. It is structural. It goes to the very foundations of how liberal markets and liberal democracies have coevolved for almost two centuries" (2006, 1). Castells' and Benkler's conclusions are shared by a group of writers who study and comment upon the influences of information and computer technology on modern society.

One of the factors most closely associated with the changes leading to an unpredictable society is the access to vast, global, and instantaneous electronic information. Unlike the print and analog electronic media, digital information can be created, copied, edited, and disseminated by any individual with modest skills and at nearly zero cost. In the landscape of digital information, individuals are more active in consumers and creators of information than they are in print-dominated landscapes.

Prosumers Realized

Whereas previous generations were primarily consumers of media, there is an emerging expectation that individuals will participate in the creation of the digital media landscape as much as they consume in that landscape. Alvin Toffler (1980) is credited with introducing the term *prosumer* to describe the pattern of media use that he accurately predicted would dominate in the digital age. The term combines producer and consumer, and a prosumer is described as one who does both, sometimes simultaneously. *Wikipedia*, the open source encyclopedia, is an example of the construction of knowledge and media by prosumers that typifies the social construction of knowledge in the digital world; a single *Wikipedia* user may be creating original content, consuming others' content, and improving others' content as he or she edits a page. The prosumer has a more sophisticated relationship with information than the consumers who came before. As instructionism is replaced as the dominant pedagogy, learners will gain experience as prosumers. Several researchers and research groups have explored the nature of the prosumer experience, and these groups are describing academic skills that will prepare students for full participation in the emerging milieu of technology-using prosumers.

In the different vocabularies used by these research groups, a similar set of skills essential for the digital generations has been articulated for educators. The collection of skills described by these scholars is not amenable to instruction. The skills are applicable to many and unpredictable situations and the skills require sustained practice to be continuously developed. Just as reading and writing and computing (all on paper) were the foundational skills that organized education for previous generations, the skill summarized in this section will be the foundational experiences for students in the future.

Mark Deuze (2006), a scholar from Indiana University, Bloomington, identified participation, remediation, and bricolage as skills needed for the 21st century media landscape. Social networking sites and media sharing sites are examples of technologies that encourage this participation. The Internet provides access to vast information from sources of dubious reliability; this necessitates individuals take a more active role evaluating information than was necessary when most print-based information had been professionally reviewed and edited. Bricolage is a term that refers to one's openness to exploring new technologies and tools, discovering both how a new tool can be used to perform familiar tasks and to discovering how a tool can be used to perform tasks not previously known. Being a bricoleur requires one to approach a new technology with openness to new connections and without feeling compelled to follow prescribed patterns of use.

Richard Mayer (2001), a psychologist form the University of California, Santa Barbara, suggested that students who are actively engaged in learning will be selecting, organizing, integrating, comparing, generalizing, and classifying information to solve a problem. Selecting information has become more complicated as students can access more and more types of information, and users must attend to the details so that they can identify the source of and the credibility of information they consume. In 21st century classrooms, organizing information requires learners to place the information in the structure of the discipline that has been included in the scaffold provided by the educator or the expert, as well as organizing to both reflect and present. Integrating information requires the learners to find connections; cognitive scientists have discovered that building connections between new information and existing understanding is necessary for learning, and the more opportunities for finding connection, the more likely the information is to be learned. Comparing information requires the learners to evaluate the information in light of other information and reconcile differences when finding connections; this is the same task that Deuze called remediation. Generalizing information requires the learner to justify conclusions based in information and to find common themes between different but similar information. Classifying requires learners to find common elements or type-information to categorize information.

Bertram Bruce and James Levin, education researchers from the University of Illinois at Urbana-Champaign, suggested several varieties of technology-rich active learning exist (1997). Inquiry is active learning that requires students to build and test theories. Testing theory requires students to access, collect, and analyze information and those are processes that can be facilitated using information technology. Active communication in computer-rich learning environments necessitates learners prepare documents to express new understanding; information technology provides many options for active communication using words and graphics in various media in both isolated and social environments. Leaders at North Carolina State University also defined similar skills when they included problem-solving, empirical inquiry, research from sources, and performance in their plans for creating a technology-rich teaching and learning in the university (Margolis 2004).

Conclusion

Teaching and learning in 20th century classrooms were largely built on behaviorist psychology and direct instruction. The roles of teachers and students were clearly defined and the curriculum was relatively consistent over generations as it was based on using print-based information. While new practices were developed for each generation, the dominant role of print as the information technology limited the extent of the changes. In the 21st century, networked digital technology necessitates consumers and producers of information be more sophisticated than they were when print dominated.

A Brief History of Computing in Schools

The history of any technology plays an important role in defining its future. The devices that exist along with the social conventions surrounding those devices become the foundation for future activities. It requires time, energy, financial, and political resources to modify the social practices and the technological landscape of a school, so extant technology contributes to future directions. For these reasons, it is necessary to briefly review the history of computing in schools.

Historians of technology trace the beginnings of computers from the analytic machine invented by Charles Babbage in the 19th century. The history of electronic digital computing is usually measured from the creation of Electronic Numerical Integrator and Computer (ENIAC), the computer built to handle the massive computations necessary for military applications (including for the Manhattan Project that designed the nuclear bombs dropped on Japan) during World War II. After the war, computers were slowly introduced to information-rich industries including those related to military production, insurance companies, and airlines. In the years immediately after World War II, annual sales of computers could be counted with single digits.

After its slow start, however, the computer industry grew rapidly with both industry push adding new models and market pull driving innovation and defining new computational needs. As the market expanded, more money became available to support research and development by computer companies. These investments led to advances that expanded the capacity of computers while reducing the costs. Both of those factors expanded potential consumer base for computers to more organizations with more diverse needs and more modest budgets. Many organizations adopted computers to manage and manipulate information. Since computers arrived, educators refreshing their practices for new generations have negotiated the role of the technology in their work.

Computers Start School

By the early 1960's the price of mainframe computers had decreased to the point where sales to educational markets were possible. At about the same time the potential of using computers in schools was recognized, but some observers have suggested that the educational applications of computing were invented so that sales could be made to that market. Regardless of what goals motivated early advocates of computers in schools, computers did begin arriving in schools within decades of their invention.

Don Bushnell writing in a monograph for the Department of Audiovisual Instruction for the National Education Association in 1963 predicted, "the digital computer and its peripheral equipment will support most of the subsystems in the total school complex" (cited in Bushnell 1964, 56). Bushnell predicted computer-rich classrooms would be places in which a standardized curriculum would be delivered to all students via computer terminals; students' learning would be measured based on their being able to provide correct answers to questions posed by the computers (the correctness of answers being judged according to those answers stored in the computer). This picture of highly standardized computer-mediated curriculum and instruction was promoted as a very efficient method of instruction, and it reflected the dominant pedagogy of the time.

In the education envisioned by Bushnell, students and teachers would interact with information, but the devices would still remain unseen by and untouched by students and teachers. This was not unlike the model of computing in other businesses and industries. Improved efficiency and productivity were predicted for all computer-mediated tasks even when technicians still operated computers and computers were programmed by physically reconfiguring the circuits. The development of the general purpose computer that sat on a desktop and came with a graphical user interface to facilitate use by almost anyone to perform almost any desired function was still decades in the future, but still advocates from information industries were predicting increased efficiency.

Computers Arrive in Classrooms

In the 1970's computers entered the consumer market, and hobbyists began purchasing computers. By 1981, personal computers could be purchased for less than $1000, and amateur enthusiasts (including children) were writing their own programs to satisfy their own interests and curiosities, but consumer computers were still marginalized and largely a hobby.

Joseph Deken, a statistician working at Stanford University who had received his first computer training in 1964 at Kansas State Teachers College, suggested that the coming decades would see individualized and decentralized teaching and learning that would be delivered through information technology. While not a computer utopian, Deken did advocate for using computers in schools because of the benefits for students. Deken opined "how much better to be free to roam about at will and speak the language with teachers as 'natives' to assist you" (1981, 247). From the perspective of the second decade of the 21st century, Deken's reference to teachers as digital natives strikes many as odd; for years, the term "digital native" has been used to describe those who were born into a world with digital devices whereas "digital immigrant" has been used to describe those who are older than digital natives and who are trying to become competent with the natives' tools. In 1981, teachers were more skilled than their students at using print, so it was natural to expect adults to always be more skilled users of the dominant information technology than students. Ours is undoubtedly the first culture to experience a technology skill inversion: younger generations are more experienced with and more confident with a dominant technology than older generations. The implications of this growing skill inversion for educators are considered in chapter three "Learners."

Other high-profile writers were advocates for using computers in schools as well. Michael Crichton (the same novelist who would later write *Jurassic Park*) wrote *Electronic Life* in 1983; the book was his response to friends and acquaintances who were constantly seeking his advice on buying, setting up, and using their first computers. His book introduced readers to the vocabulary related to and ideas about computers that, he believed, would become familiar as computers became embedded in everyday life. Crichton saw computers as holding great potential for encouraging creativity, and he observed, "One of the great delights of any new technology is that it is for a while, free" (1983, 28). He continued, however, to describe how a group that he called computer Calvinists were—even early in the history of computing for the general public—already at work to ensure that computer use became standardized and that users learned and followed the rules. Crichton expressed his desires for computers in schools and throughout society with these words:

> Personally, I hope that, for once in the 20th century, a technology stays free. Because the rules-makers always manage to kill the essence while tidying up the details. Dogma replaces direct experience, and ritual becomes reality (28).

Unfortunately from Crichton's perspective, the computer Calvinists appear to have exerted their influence over computers in schools.

In 1994, Seymour Papert, the mathematician from the Massachusetts Institute of Technology who was a pioneer in using computer programming to teach mathematics to young children, suggested the history of computers in schools could be deconstructed into three phases (see table 1.2). First, there was a brief time when innovative educators had computers in their classrooms and engaged students with them (one gets the sense that computers in those classrooms were free in the manner Crichton hoped). Second, there was a period during which computers were centralized in computer labs and specialists assumed responsibility for teaching computer classes; computing became a subject matter.

This was similar to the centralized computing and data predicted by Bushnell in 1963 but was less democratic than Bushnell had predicted, because computer classes were primarily filled with rich white males. Papert argued that the early innovators had been right with their approach, and he encouraged educators to initiate the third phase of educational computing which would find them returning to the innovative curriculum and practices that characterized the initial endeavors with computers in classrooms.

In the years since Papert wrote, educators appear to have taken his advice, at least partly. In many schools, computers have returned to classrooms, and they are more likely to be used to support teaching and learning in those places rather than in special rooms. Laptops are common in schools (including Chromebooks which are Internet-only laptops) and in some communities students are encouraged to use their own laptops and tablets in class. In these schools, technology is often said to be "integrated" into the classroom, although there is little agreement about what comprises appropriate integration.

Table 1.2 Three phases of computer placement (adapted from Papert 1994)

Computers arrive in classrooms	Computers placed in computer rooms	Computers return to classrooms
Few machines (1 or 2 for the entire class)	Many computers (one for each student)	Many computers (one for each student)
Diverse instruction	Common instruction	Integration into curriculum
Managed by teachers	Managed by specialists	Managed by specialists

The theme that schools are technology-rich places, but that education has not changed in response to the technology, is common in the educational and the popular literature. In the middle of the first decade of the 21st century, two scholars, William Pflaum (2004) and Ellen Seiter (2005) spent time observing and documenting the computer use in diverse schools. Both found that word processing, spreadsheets, and drill-and-practice software were among the most common uses of computers and information technology in school. By then schools had been connected to the Internet for several years, but the network was mostly observed to be distracting students from their lessons. Both Pflaum and Seiter concluded that computers appeared to play a minor role in K-12 education, and many uses were diversionary rather than instructive. Clayton Christensen, a Harvard business scholar known for his work with innovation, along with colleagues Michael Horn and Curtis Johnson recognized the largely successful efforts to obtain and install and connect computers, but concluded that "classrooms look largely the same as they did before the personal computer revolution, and the teaching and learning process are similar to what they were in the days before computers" (Christensen, Horn, and Johnson 2008, 70).

Educational researchers whose work has extended into the 21st century have made similar observations. Matthew Pearson and Bridget Somekh (2006), educational technology scholars from the United Kingdom, observed that advocates for using computers is schools retain the instructionist approach, as it is perceived to be a method for "improving the efficiency of schools and the impact of teachers' 'whole class' expository" and also for "motivating learners and ensuring that schools keep up to date with the information society," they conclude, however, that "it is rare for alternative constructions of ICT to be posited" (531). (They use ICT to abbreviate information and computer technology as previously cited authors.) Pearson and Somekh's alternative constructions include practices and approaches to teaching and learning that are aligned with deeper learning principles.

The conclusions of both educational scholars and observers of education appear to be converging: networked digital information technology is available in schools, but teaching and learning is unchanged. Computers appear to have simply substituted for previous technologies, and have not changed how teachers teach and how students learn. Much of the technology-based instruction that has been common in schools classrooms since it arrived in the 1980's can be categorized as one of four models that are differentiated by the prepositions about, by, with, or via. Activities that can be described with each of these prepositions are included in "technology integration."

Teaching *about* computers. One of the early models of computer curriculum was referred to as computer literacy, and it was based on the assumption that students who learned the parts and functions of computers systems would be "literate," thus able to use computers for practical purposes. While this model fell out of favor quickly, the idea that students must be taught about computers remains. Efforts to include Internet safety and efforts to introduce computer programming to students can be understood as teaching about computers as the lessons have little relevance if technology is not the focus.

Teaching *by* computers. Games designed to teach, which are commonly sold as "edutainment," are an excellent example of teaching by computers. The pedagogy is simple and instructionist: Present information to users, and then test their understanding by posing questions. If the user gives the correct answer, then give positive feedback in the form of audio and visual rewards. As high-stakes testing has gained importance in evaluating schools, many vendors sell test preparation software that follows this model. Using sophisticated algorithms, those programs track student answers and present different information based on responses, and all can proceed with little or no intervention from the teacher.

Teaching *with* computers gained popularity under the term "technology integration" around the turn of the century. David Jonassen (2000), a scholar from the University of Missouri and a leading advocate of technology integration, argued that in classrooms where technology has been integrated computers will "necessarily [engage] learners in critical thinking about topics they are studying, which, in turn, results in better comprehension of the topics and the acquisition of useful learning skills" (22). Jonassen's term has been applied to all screen time for students, despite the original application for teaching and learning in which computers and technology was a tool just as books and pencils had been tools previously.

Teaching *via* computers is illustrated by the very popular "PowerPoint and projector" lecturing that one observes when walking around many schools today. With this model, the computer has replaced the overhead projector or the chalkboard of previous generations. Another common example is the word processor, which is a replacement for the typewriter or other writing tools. In each case the classroom dynamics are unchanged, but—ostensibly—technology appears central to the teaching.

More recently, *flipped classrooms* have been presented as a model of technology-based education. In general, the model finds teachers pointing students to instructional videos that are available on the Internet. Students watch those videos as a substitute for a teachers' lecture, and then the work that would have been assigned for homework is completed in class. The advantages of this model are that students can watch the video as needed so the lesson can be reinforced and also that students complete follow up work with the teacher available to give further guidance and support. While those are reasons for adopting the model, critics point out that this is really simply instruction by computers and the limits of all instructionist pedagogy can be applied to flipped classrooms.

Mission Accomplished?

The computer technologies introduced into schools in the last quarter of the 20th century were part of a long parade of electronic information technologies that were introduced first into the popular culture and then into education throughout the 20th century. In 1986, Larry Cuban, a professor of education at Stanford University, reviewed the history of radio, movies, and television and he observed a common pattern describing how the technologies failed to transform education as predicted and eventually fell into disuse in schools. First, advocates argued that each technology could be used to make teaching more efficient and more effective. Second, dubious research (frequently supported by the manufacturers of the technologies) was used to support the advocates' claims. Third, the technologies were introduced to schools, but soon fell into disuse. Finally, the next technology with promise to transform teaching was introduced and the pattern was repeated (see figure 1.2).

Cuban identified several reasons for this pattern. First, limited access to the technology posed an obstacle to its use. The expense of obtaining equipment, the need for expertise to maintain and operate the equipment, and the inflexibility of schedules were all factors that limited teachers' access to electronic information technologies throughout the 20th century. Access was also limited by teachers' inability to operate the equipment, and inadequate training exacerbated this factor. Second, a lack of curriculum materials dissuaded teachers from using these technologies; because other resources were available that were better-aligned with their curriculum and with their previous teaching experiences, many educators avoided the new technologies preferring to use the safe, familiar, and reliable materials which they knew. Finally, many decisions to introduce these technologies to classrooms were made by leaders who had little understanding of the logistical challenges that inadequate access and training and weak curriculum materials posed for teachers.

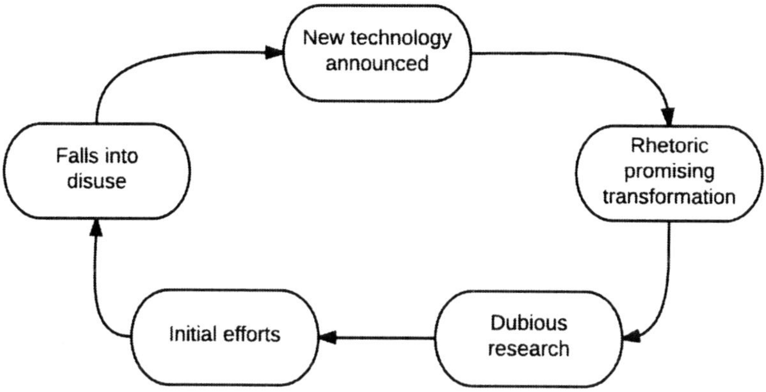

Figure 1.2. Pattern of technology implementation (adapted from Cuban 1986)

When introducing computers to classrooms, school and technology leaders appear to have paid attention to the lessons learned by the failure of radio, movies, and television to deliver the promised effect on teaching. There have been systemic efforts to ensure all schools have the resources to install computers and connect them to the Internet; there have been efforts to develop resources to support curriculum planning, and there have been efforts to provide educators with on-going professional development. While they added computer processing power and memory, broadband Internet connections, and useful software to their schools, school and leaders also hired professionals to manage computer and information systems and others to work with teachers to support technology-based curriculum and instruction. Today, it is not unusual to find schools employing computer technicians, network administrators, and technology coordinators in addition to technology integration specialists and technology teachers.

Connected Schools

Once the World Wide Web became available in the mid-1990's, the Internet changed from being a resource for academic and government researchers to being a tool for commerce and the people. Many educators recognized the

World Wide Web as an opportunity for students to access previously unavailable resources. Because few schools had the network infrastructure necessary to connect classrooms to the Internet in the mid-1990's, and because few schools had the necessary budget to install that infrastructure, volunteers in a large number of communities organized NetDays, events during which community members installed network cables in school buildings. Through active participation in such grassroots efforts, citizens demonstrated an intense interest in ensuring the Internet was available in schools. Such interest was identified as one factor that led the officials in the administration of President Clinton to propose several programs to improve technology capacity in K-12 schools in the United States.

The passage of the Telecommunications Act of 1996 established a program through which schools and libraries were reimbursed for a portion of the costs of telecommunications service, with deeper discounts going to communities with larger economically disadvantaged populations. The program came to be known as e-Rate and currently it is administered by the Universal Service Administrative Company, and the money it provides continues to support access to voice and data networks in K-12 schools. At about the same time the e-Rate program started, the federal government in the United States initiated a program called Technology Literacy Challenge (TLC) that made government funds available to purchase computers for schools. While TLC has been discontinued, and federal support for computers in schools is variable, local efforts to increase students' access to computers continue to be popular. In many communities, those efforts to provide each student with a computer are known as one-to-one initiatives.

Data from the National Center for Educational Statistics indicate that in 2005, 94% of the instructional spaces in the United States had access to the Internet; for some populations (for example in urban schools) the percent was as low as 88%, but for other populations the percent exceeded 95% (Snyder and Dillow 2013). The same report indicates that in 2008 for all schools serving reported subpopulations (elementary and secondary; urban, rural, and suburban; small, medium, and large; all quartiles of students receiving free and reduced lunch) the ratio of students to computers in schools was at least 3.4 for subpopulations, and 3.1 for all schools together. Of these computers, 97% had Internet access

Curriculum Standards

One of the reasons that educators struggled to find the appropriate model of technology-rich instruction was the novel challenge presented to educators by new and unfamiliar tools. Around the turn of the century, the International Society for Technology in Education first published the *National Educational Technology Standards for Students* (NETS-S), and those became an organizing structure for the design of technology-rich curriculum and instruction in all classrooms (NETS Project and Brooks-Young 2007). Concurrently, authors such as David Jonassen (1999 and 2000) and Janet Ward Schofield (1995) were arguing that how technology was used in the classroom was as important as the nature of the technology that was available in determining the effects of computers on students' learning.

While the ISTE standards encourage computer-based experiences for students focused by a) creativity and innovation, b) communication and collaboration, c) research and information fluency, d) critical thinking, problem solving, and decision making, e) digital citizenship, and f) technology operations and concepts; educators frequently describe pressures to use screen time for test preparation and other forms of instruction. This has led to an on-going dialogue within communities of educators about the appropriate role of computers in the classroom, and the appropriate lessons to teach about devices.

Teacher Training

Judith Sandholtz, Cathy Ringstaff, and David Dwyer (1997), along with Janet Ward Schofield (1995) are scholars who participated in the Apple Classrooms of Tomorrow (ACOT), which is widely recognized as the first effort to put computers into classrooms and to study factors influencing how they were used. One discovery traced to that project is that teacher training is an essential condition for technology to be used. Those scholars identified basic computer operation, using technology to support instruction, and managing technology as necessary for a complete professional development plan. Since then, many government grant programs and other alternative funding organizations require a portion of the funds for educational technology be used for professional development. Further, many jurisdictions require teachers to participate in continued learning about educational technology as a condition of employment or for continued certification as a professional licensed educator.

In recent decades, many educator-licensing bodies have defined a new certification area. Given various names, usually a combination of technology and integration and specialist (or related nouns), this person is recognized as an educator with greater than usual expertise using technology in the classroom. This professional provides many options for teacher training, including team teaching and other *in situ* experiences.

Conclusion

It is reasonable to conclude that, as the 20th century ended, school and technology leaders had proceeded in a manner to avoid the difficulties encountered by those who attempted to transform education using radio, movie and television. Students attended class in buildings with computers and network connections, their teachers had curriculum materials and they had access to training (or at least curriculum guides had been published and the need to provide teacher training was recognized (the degree to which individuals used those resources and accessed that training can be debated). Despite all of these efforts, many continue to observe classrooms have largely been unaffected by computers.

For more than 30 years, computers have been a factor influencing how educators refresh their practices. Books and periodicals (enough to fill libraries) have been written, careers have been built around, and billions of dollars have been spent on efforts to transform curriculum and instruction so that teaching and learning in our society's classrooms reflect the economic, political, and social realities of our technology-rich culture. At this point, early in the second decade of the 21st century, it is reasonable to conclude these efforts have largely been unsuccessful. In most classrooms, the pedagogy developed by and for 20th century society is common despite the arrival of 21st century digital technologies. It appears the refreshing of education Bruner claimed accompanies each generation has skipped the current generation.

Some individuals and groups will characterize my conclusion that classrooms have been unchanged by computer as uninformed, and they will be able to point to initiatives in local schools through which classrooms have been transformed by digital technologies. If the technology supports and maintains the instructionist models that have been described, however, we must conclude there has been no transformation.

The school reform and improvement efforts of the last few decades have advocated superficial pedagogical changes and leave unaddressed the underlying assumptions about human learning, human society and the role of technology in those phenomena that are the foundation of schooling. Evidence from several fields is challenging those assumptions that supported 20th century education, and educators are only slowly recognizing the differences between old assumptions and new ideas. In this book, I seek to identify a rationale for revising all aspects of technology-rich education, and identify general guidelines for classroom redesign and models for initiating those efforts. Most of this book addresses strategies for understanding teaching and learning in the new technology landscape, but first, I present a lens for understanding school planning.

Chapter 2
Education is a Wicked Technology

In general, one can conclude there is consensus regarding the role of formal educational systems: the public supports and maintains the system to prepare youngsters to participate in the economic, political, and cultural life of society. The nature of the experiences designed to meet this purpose changes over time. Traditionally, the domain of education incudes a wide range of subjects such as literacy (including reading and writing), numeracy (the mathematical equivalents of reading and writing), specific subject area knowledge (sufficient to become an informed citizen), acculturation to include civics (sufficient to permit participation in our democracy), vocational skills (to contribute to the economy), personal skills (to support healthy development), and academic skills (frequently called critical thinking, problem solving, or lifelong learning).

To accommodate these goals, the curriculum in K-12 schools has expanded in recent decades so that it now includes topics such as advanced mathematics including computer programming, a broad survey of the sciences including the social sciences, foreign languages, performing arts, visual arts, physical education, health, and the trades. Despite evidence that experiences in the arts and opportunities for physical activity are associated with higher levels of academic performance, it is not unusual for those activities to be perceived as superfluous to the academic or intellectual purposes of education, thus these tend to be cut from school budgets in misguided attempts to improve academic performance.

While individuals will identify all of these goals as purposes of education at any moment, changing conditions or situations cause individuals and groups to update their priorities regarding these. The global recession that occurred late in the first decade of the 21st century and that continues to influence economic conditions in the second decade refocused attention on the economic advantages of becoming educated. In particular, political leaders are recommending students focus on science, technology, engineering, and mathematics as these are anticipated to have the greatest economic effect. The more insightful leaders include the arts in that list.

Debates about the appropriateness of diverse curricular offerings illustrate the divergent opinions about the purposes of education within a jurisdiction. Differences arise from perceived economic demands of providing a diverse curriculum, advantages for students who experience the curriculum, concerns about the moral or ethical implications of some curricula, and concerns about the adequacy of the skills students develop in schools. In the Unites States, also, laws are in place to prevent disadvantaged sub-populations from being discriminated against so as to limit their access to public services, including education. All of these factors complicate the task of reaching consensus on the purpose of school or even defining the essential characteristics of schools with clarity.

In an environment of many and conflicting points of view regarding school expectations, it is reasonable for school leaders to undertake strategic planning on a regular basis. The intent of this activity is to promote school organization and pedagogy that results in the purposes of the school being accomplished for the populations it serves. Achieving these goals (as defined and measured with these plans) in large part defines the activity within a school.

In the first chapter, instructionism was presented as a vestige of education that is no longer supported by the discoveries of the learning sciences and the information landscape of the technology-rich society. In this chapter, the lens of wicked problems is presented and this lens clarifies planning that is appropriate for schools.

WICKED PROBLEMS

For the last several decades, school planning has focused on first setting goals or defining expected outcomes, then designing and implementing systems to accomplish those goals, and finally evaluating the success of the system. The pattern then is repeated with new goals defined from the conclusions reached during the previous evaluations. In this, educators are following the strategic and logistic planning that has been common for leaders of other organizations. For the generations of educators prepared and practicing in the recent decades, defining objectives is deeply embedded in their school planning strategy

This approach to educational planning is exemplified by the approach of defining educational outcomes according to scores on standardized tests, which are aligned with the standards. (As we have seen, this approach is dubious. Which tests, which standards, which actions, and the definition and validity of the alignment are subject to quite vehement and contentious debates.) Leaders advocating this approach often confuse correlation with causation; they assume actions for which they are responsible caused any improvements in performance on the tests. These claims ignore factors such as regression to the mean and they fail to validate the findings of test data with other data sources in a formal manner.

Planning: A Brief History

The origins of goals-based or objectives-based or standards-driven planning can be traced to the middle of the 20th century. In 1960, Charles Hitch, writing for the RAND Corporation first identified the role of goals in planning and noted, "We must learn to look at our objectives as critically and as professionally as we look at our models and other inputs" (cited in Rittel and Webber 1973, 155). In the same decade, initiatives undertaken by leading professional organizations adopted a similar stance, so planning in many areas of business and public policy began with questions such as "What should our systems do?" and "What are the desired outcomes for our organization?" According to Hitch, defining the goals of a system and measuring the extent to which they were achieved is as important to designing and developing accurate, complete, and efficient systems developing and refining the procedures that define system operation.

Hitch was reacting to the observation that methods designed to improve the efficiency of systems, which had solved many of the problems in science and engineering that emerged during the industrial age, were increasingly less effective. Problems had emerged that were very complex and increases in efficiency no longer produced desired outcomes or increases in efficiency have introduced new and unexpected problems.

Lars Skyttner, a scholar from Sweden, suggested in his 2006 book *General Systems Theory* that problems such as environmental degradation, artificial intelligence, and technologies of war had emerged in the 20th century that could not be solved through improved efficiency. It was reasoned that such problems are difficult or impossible to solve or even to understand using the reductionist methods employed by natural scientists, and he observed,

> Interactions of systems-variables are so interlinked to each other that cause and effect is a kind of circular logic. One separate variable thus can be both cause and effect. An attempt to reduce complexities to their constituents and build an understanding of the wholeness though knowledge of its parts is no longer valid. (37).

Both Hitch and Skyttner recognized the increasing complexity of problems, and each introduced more sophisticated problem solving than was common at the time. Complex and irreducible problems with a social dimension (such as education) are even more complicated and difficult to solve. Since the 1970's these have been recognized as wicked problems.

Identifying Wicked Problems

In the 1973 article, "Dilemmas in a General Theory of Planning," Horst Rittel, who was a professor of the science of design at the University of California, Berkeley and Melvin Webber, who was a professor of city planning at the same institution, differentiated tame and wicked problems. In many areas of human endeavor, problems are definable, understandable and consensual. Each problem can be attributed to one (or a small number of) clearly identifiable and definable cause(s), the problem can be known through systematic study, and there is agreement when it has been solved satisfactorily. Rittel and Webber used the term *tame* to capture the nature of those problems. Tame did not mean the problems were simple to solve or that the problems were not important and ethically challenging. Tame did mean that the problem can be studied and solved through science and engineering. While recognizing that no problem can be completely tame, Rittel and Webber suggested that the living conditions for large populations of humans had been improved by solving tame problems related to (for example) health care and medical procedures, transportation and sanitation systems in cities, and agriculture.

Rittel and Webber concluded, however, that most social problems (such as education, law enforcement, and local governance) are not tame, and they introduced *wicked* to capture the nature of the problems. The causes of wicked problems cannot be clearly defined, they cannot be easily understood, and the resolution (or even the existence) of the problem depends on one's interpretation. Solutions to these problems must be tentative and qualified. Rittel and Webber argued the methods for solving tame problems will not produce adequate solutions for wicked problems, and even defining a wicked problem as if it were tame is an impediment to planning solutions that can be deemed sufficient.

Characteristics of Wicked Problems

In their original article, Rittel and Webber defined 10 characteristics of wicked problems. Other scholars have reduced the number of characteristics by combining some that are similar. Regardless of the number of characteristics used to define wicked problems, scholars who focus on these problems concur wickedness arises from the social nature of the problem and the diversity of valid and mutually contradictory perspectives on the problem and the solution that can be supported with both reason and evidence.

Ill-defined

There is no definitive formulation for wicked problems. In the natural sciences, problems are clearly and completely defined, and all who are familiar with the field will agree on the nature of the questions and the nature of the methods that should be employed to solve the problem. (This holds as long as the research works within the dominant paradigm of the field.) If a tame research problem is too complex or too multifaceted, the scientist defines the problem with greater precision and takes steps to control all relevant factors. Such control over the boundaries of and within a problem is not possible when solving wicked problems; the boundaries between causes and effects blur and unidentified causes may be influencing the resolution of the problem for the entire population or for a part of the population. In many instances, the result is the wicked problem and its solution become inseparable. The decision to implement a different solution would have been a decision to solve a different problem or to address a different factor affecting the wicked problem.

In a similar way, there are no rules for guiding or limiting planners as they design solutions to a wicked problem; any solution is permissible. In many cases, the strategies depend on how the problem is framed and what the anticipated solution is. Further, no rule objectively determines when a wicked problem has been solved. When assessing a solution to a wicked problem, planners frequently find unresolved sub-problems, unrecognized causes, or new problems. Whereas evaluation of tame solutions depends on measuring the extent to which goals were accomplished, the evaluation of wicked problems typically leads to deeper understanding of the problem or recognition of previously unknown relevant factors.

Rather than being judged as accomplished or not (in a true-false manner using an objective and pre-defined measure), solutions to wicked problems are judged as better or worse in a subjective manner by those who experience the solution. The judgments are made according to (probably changing) circumstances at the moment the judgment is made. Adjectives such as good or bad, working or dysfunctional, acceptable or not acceptable, satisfactory or unsatisfactory are common when people assess solutions to wicked problems, and these terms demonstrate the necessarily subjective assessments. These adjectives are generally not acceptable to those who seek to solve tame problems in an objective manner. Equally unacceptable to those looking for tame evaluations of wicked problems are answers to the question, "Did we meet the goal?" with answers that begin "It depends…." These adjectives and answers can (which are necessary when evaluating the solutions to wicked problems) cause school leaders to react with consternation when there is discordance between their assessment and others' assessment of solutions that are implemented. They are expecting tame answers to wicked problems.

Every Solution is Experienced

Tame problems are generally solved through processes that are tolerant of error. Engineers typically build and test models and prototypes before any solution is fully deployed. Through this process, flaws in the design are identified and mitigated prior to releasing the design for use. (Examples of engineering failures are well known, but given the number of engineered products that are used on a daily basis, the number of failures with catastrophic consequences is remarkably low.) Planners solving wicked problems do not have the option to test their solutions in the manner scientists and engineers do.

Because they are social, solutions to wicked problems are only implemented when experienced by humans. This experience will have permanent effect on those humans. For example, educators who are preparing a mathematics curriculum cannot implement it without introducing it to students. That exposure will affect how students understand mathematics. Experiencing the curriculum may improve students' abilities to solve problems and learn more advanced mathematics later, or the instruction may inhibit their abilities. The experience may also change students' attitudes towards and beliefs about mathematics and their skill in using mathematics. A curriculum that is poorly developed or implemented may result in a group of students who are strong math to become confused or to believe they are weak students because they cannot learn the material.

Each Wicked Problem and Solution is Unique

Wicked problems tend to be local and interconnected. The exact nature of the problem and its causes in one community is different from the similar problem in another community. As a result, the understanding of the problem is different and the judgments of solutions will be different as well. Because of this, attempts to transfer solutions from one community to another often do not produce similar results because important conditions of the original setting are not replicated in the new settings.

Wicked problems exist at multiple levels, and each wicked problem is the cause of another (or several other) wicked problems. For those who seek to understand a wicked problem and create and implement a solution to it, this complicates the task by making it impossible to identify and control all relevant factors. Even if key causes can be identified, those causes may arise from a source over which the problem solver has no control. Further, a sufficient solution that can be implemented may cause unacceptable problems to arise for others. As a result, even the best made plans may be abandoned or prevented from working because of a cause that cannot be controlled by the decision makers or that will produce and an effect judged unacceptable by others who are politically more powerful.

While it may not be possible to transfer a wicked solution developed in one setting to another, it is possible to mix and remix solutions designed in one setting when designing solutions for another. This characteristic of wicked problems necessitates planners demonstrate innovative and creative thinking when designing solutions; simply following the recipe developed by another planner is a dubious strategy for planners of solutions to wicked problems, but adapting and exapting others' solutions is a viable approach.

For educators, this character of wicked problems suggests that the methods designed for one classroom and evaluated as effective in that classroom are unlikely to produce the same outcomes in another classroom. Factors ranging from the previous experiences of the students to the skills of the teacher to the sociocultural context of the communities in which the school is situated all influence how a curriculum and instruction problem and solution is understood and instantiated in a community.

TECHNOLOGIES

In general, humans like to categorize using dichotomies; an object is placed in one group or another. As an undergraduate student preparing to be a science teacher, I classified plants using dichotomous keys in botany class, hours examining specimens (usually alone and while drinking coffee) to decide if a specimen demonstrated a trait or not. At the same time, I was a student in education. Classmates and I debated various questions about teaching and learning, hours discussing (with others and while drinking coffee) whether education is a science or an art (for example). Rather than being merely a speculative conversation among undergraduates, how one answers the question "Is education a science or an art?" has important implications for how one approaches the process of education and how one seeks to improve education.

Education-as-science approaches curriculum and instruction as a tame problem. According to this view, teaching can be reduced to taking well-established and highly controllable actions and the effects of those actions can be measured with reliability and validity (the same actions can produce the same result when repeated with other populations in similar settings and the action does actually cause the effect that is measured). Advocates of education-as-science also tend to predict that the same pedagogy can be used with equal effectiveness for all students and in all curriculum areas.

Those who argue that education is an art typically hold that the reduction of teaching to well-known causes and effect relationships is not possible, and that effective education cannot be described as well-established principles and only through observation and insight (especially into differences between individuals) can one be an effective educator. Further, those who believe education is an art hold that it is difficult to measure learning and that the factors influencing effective teaching are difficult to deconstruct and control. Although this debate can sometimes lead to a deeper understanding of the characteristics of education for all stakeholders, it does suggest that education is either one or the other; but that is a false dichotomy. In the 21st century, a third option must be recognized: education is a technology.

Technology: A Broad View

Historians of technology recognize two types. Hard technologies include the artifacts—from stone axes to automobiles to computers—built by humans. Soft technologies include those practices—from language to banking to computer software—that function as technologies but that cannot actually be held. The formal education systems familiar to students approximately aged six through 18 are an example of a soft technology; it demonstrates the characteristics of all technologies, but it cannot be held. By understanding the educational system as a technology, educators can better explain the causes and effects of various factors on and in their classroom than if they consider education to be either a science or an art.

W. Brian Arthur, an economist and complexity scientist, asserts, "Technology in fact is one of the most completely known parts of human experience. Yet of its essence—the deepest nature of its being—we know very little" (2009, 13). Despite humans' rich experience with technology, Arthur observed, "we have no agreement on what the word 'technology' means, no overall theory of how technologies come into being, no deep understanding of what 'innovation' consists of, and no theory of the evolution of technology" (2009, 13). Theory represents scholars' best knowledge of how the field actually works. Theory helps scholars determine which problems need study, the design of methods to study those problems, and it gives guidance for interpreting the results of their studies. Any field that lacks a unifying theory is likely to be characterized by frequently changing and mutually contradictory ideas being advocated by experts.

The lack of a general theory of technology is an observation that can be applied to 21st century education. Educators have detailed histories of how particular approaches to education originated and developed. Educators have detailed guides providing advice on the design classrooms for specific purposes, and how to design schools to support a wide range of educational practices. Educators have deep philosophical ramblings on the meaning of education in our society, and both utopian and dystopian views of the future of education and schooling (those views usually determined by our choice to accept or reject the writer's stance). The multiplicity of the pedagogical models being proposed for teachers is evidence that the experts are each working within a different paradigm—some advocate education as science, some education as art.

Education as technology is a lens that can begin to unify these diverse and changing and contradictory views of education and the role of technology in education. Education does share the characteristics of other technologies and those characteristics appear to describe how education originates and evolves and how innovations emerge and spread through the field just as is observed in other technologies.

Characteristics of Technologies

There are several generalizations that can be made about technologies, and these define features that are common to all technologies and provide a structure with which we can explain what is observed and predict what we expect to observe. Technologies apply natural phenomena to meet a need defined by humans, technologies have a module and recursive structure, technologies are non-neutral, and the transfer of technologies is complicated.

Technologies and Human Need

All technology exists because of the intervention and invention of humans. Technologies have extended humans' bodies to allow for the physical modification of the environment on a wide scale; and when combined with humans' social nature, technologies are the foundation for the creation of culture.

The connections between humans and their technologies are deep and inseparable. In his book *The Artificial Ape*, archaeologist Timothy Taylor observed, "Key aspects of our biology would be impossibly dysfunctional without technological support" (2010, 27). It is only though our technology that our species with its weak muscles, thin skin, and small teeth (all deficits that make human survival dubious in any ecosystem) could have adapted to survive in any environment. The baby sling adds extra arms so that a mother can hold her child and still work. The stone blade allows the human hand to break what otherwise would be unbreakable (rocks, hard-packed clay, nuts, or an enemy's skull). Cooking extends digestion outside our bodies. The human species appears to extend beyond our biological bodies. Taylor concluded of technologies, "These are not the same as inanimate, natural things. They are artificial and form the nonbiological aspect of the artificial ape" (194).

Kevin Kelly, an influential writer and thinker about technology, appeared to concur with Taylor when he concluded humanity is characterized by the technium, which includes both humans' biological character and the technology they create into a single complex that "extends beyond shiny hardware to include culture, art, social institutions, and intellectual creations of all types" (2010, 11-12). This may lead to the conclusion that humans define a need and then create a technology to fulfill that need. David Nye, a historian of technology from Denmark, observed that "the central purpose of technologies has not been to provide necessities, such as food and shelter, for humans had achieved these goals very early in their existence" (2006, 2). Nye and others find that the adage "necessity is the mother of invention" is just the opposite of what we observe in humans' use of technology; as humans invent technologies and they become embedded in culture, they redefine what is necessary.

The familiar algorithm for finding sums taught to schoolchildren for generations (see figure 2.1) is an example of a technology that is perceived by many to be a necessary skill; it is deeply embedded in our culture. Coincident with the emergence of agrarian lifestyles and trade between separated populations was the emergence of accounting systems (the first writing systems) for accurately counting large numbers. Until trade emerged, it was unnecessary for humans to find large sums accurately. In cultures without writing, quantities tend to be labeled as "one," "two," and "many," and bargaining and bartering are the main methods for economic transaction (Hobart and Schiffman 1998). Even in our trade-based economy, finding large sums accurately using the algorithm is unnecessary in many situations. When one only needs an estimate of the answer, when one has a calculator (an abacus for example), when one has finely tuned abilities to find sums through mental calculations, or when one knows another

$$\begin{array}{r} 435 \\ +\ 628 \\ \hline 1063 \end{array}$$

Figure 2.1. An algorithm for finding sums

algorithm for finding sums, the necessity of that algorithm evaporates. Still, the algorithm is perceived to be a necessary part of schooling.

Mobile smart phones were unnecessary until they were invented (and became inexpensive); until the 21st century most humans survived and communicated in a manner they deemed sufficient. Now they are deeply embedded in our communication habits for many people (including me); it is necessary to contact me through my cell phone if you want to communicate with me in a timely manner. It appears that the technologies that have been invented and those that are accepted and expected in culture are among the most important factors in defining need.

The human need to prepare young people to participate in social systems has motivated the design of formal and informal education systems throughout human history. The social system for which the learner is being prepared determines in large part what comprises the education. As print dominated in the industrial and information ages, formal systems of teaching reading, writing, and arithmetic (and other fields mentioned at the beginning of this chapter) were developed. Learning, of course, occurs in many other settings as well. Children learn to cook and garden and hunt (for example) in informal settings. Apprenticeships, which are particularly common in cultures that lack writing, incorporate elements of both formal and informal educational systems, and many educators are rediscovering the advantages of content-rich education as provided by apprenticeships.

Technology and Nature

Every technology is based on a natural phenomenon. In some technologies, it is easy to identify the natural phenomenon that is either applied or controlled in the technology. The control of fire is the basis of internal combustion engines, for example. Education is a technology based on the functioning of the human brain.

Because technologies are improved, largely through greater understanding of the details of the natural phenomena and how those can be best controlled and applied to the problems, technologies tend to follow an s-curve (see figure 2.2). Performance is poor (for example it may be slow or inefficient or weak) when it is first developed, but then performance improves quickly. As the limits of the natural phenomena are approached, there is a second slowing of improvements. Often this second slowing is followed by the development of a new technology; the new technology is discontinuous from the first and begins its own s-curve pattern of development (see figure 2.3).

The transition from one technology to another demonstrates the connection between science and technology. Cyril Stanley Smith, a metallurgist who worked on the Manhattan Project, suggested, "A new thing of any kind whatsoever begins as a local anomaly, a region of misfit within the pre-existing structure" (cited in Rhodes 1999, 331). Smith's anomalies lead scientists to define research problems that lead to new discoveries of natural phenomena that

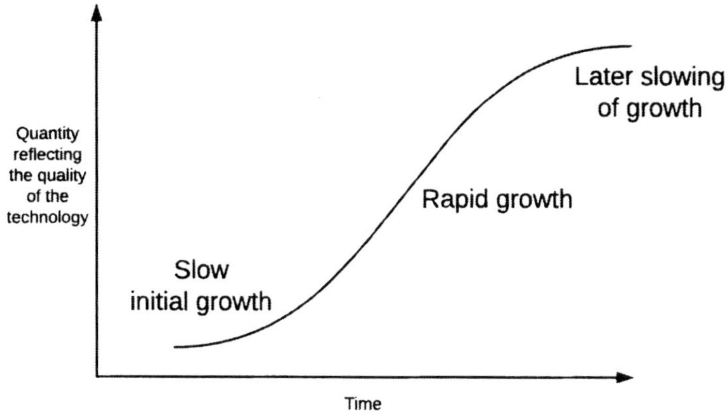

Figure 2.2. A generalized s-curve

can improve technologies; they also lead to reorganization of existing technologies to reconcile the anomaly. Anomalies motivating new technologies can arise from both observation of limits in nature (or our understanding of nature) and from changing social conditions. In education, we see example of both.

Educators who are adopting strategies aligned with deeper learning are reacting the anomaly between instructional methods and the nature of human learning. Discoveries in the learning sciences pointed to the limits of our understanding of human brain function as it was informed by behaviorist psychology. Once educators' practice reflects deeper learning, they overcome the limits of instructionism and student learning improves. Improvement of performance is redefined according to the principles of deeper learning.

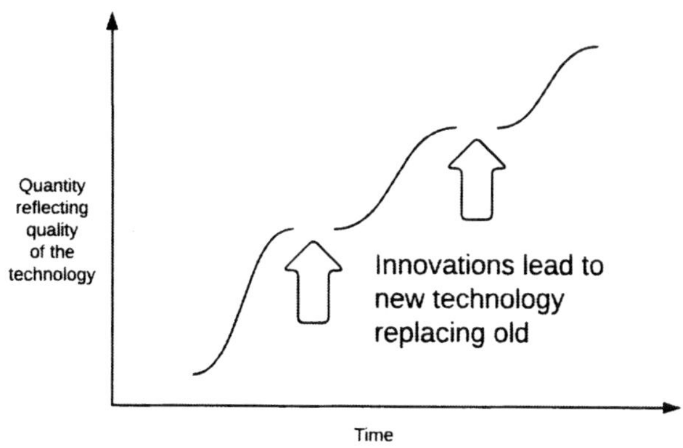

Figure 2.3. Technology replacement

In the 18th and 19th centuries, each school was organized to meet the needs of a small community and to align with the social norms and economic realities of the community. For example, the school calendar was designed to accommodate the labor needs of an agrarian society and the schools were located in communities so that students who were enrolled could live at home, educators tended to be generalists (a single teacher taught all content), and students left school about the time they reached puberty.

As populations moved from rural areas to urban areas, and as more technical skills were necessary, the high school grades with specialized teachers were added to ensure students were exposed to the skills and knowledge necessary to participate in the industrialized economy. At about the same time, schools were organized to facilitate the acculturation of large populations of immigrants. In the United States science and math curriculum and instruction was updated in response to the launching of *Sputnik* by the Soviet Union. These transitions in education, each representable by s-curves, were driven by changes in the social conditions.

Technologies are Recursive and Modular

Every technology comprises other technologies. As a result, any technology can be deconstructed into its component technologies, and each technology can be understood as part of a larger complex of technologies. Scholars refer to systems that occur at multiple levels in this manner as recursive. Although related and connected, the component technologies can be considered separate, and these modules can be treated independently of each other.

The modular and recursive nature of education is clearly demonstrated in several familiar characteristics of public schools. Physically, schools comprise classrooms that are relatively similar in nature. The typical management structure of schools also demonstrates modular and recursive organization. Large schools tend to be organized into departments, grade levels, or teams, and schools are part of larger districts; so a school comprises smaller units of management as well as being parts of larger units of management. Curriculum is recursively organized into units, and units are recursively organized into lessons.

The modular and recursive nature of technology leads to three observations that can be made about all technologies. First, modules do not exist in isolation. Because each technology comprises others and because each technology is part of a larger complex, any change in one will affect the others. Even through each sub-system of a technology can be treated as a module, technologists recognize that any changes made will affect the entire technology and complex when it is implemented. For educators, it is important to recognize that changes in one part of the school structure will have consequences for the entire system. For example, there is a connection between fitness and brain function, so changes in opportunities for recess or physical education will affect performance in all academic areas.

Second, technologies exist at many levels, so they must be understood and evaluated at many levels. Patterns and trends observed and relevant at one level may have little or no meaning at other levels, and interesting properties that are apparent at one level may not be observable or predictable from what is known about another (these are commonly called emergent properties). In education, the multilevel characteristic of technologies can be illustrated by considering standardized test scores. While those scores may illustrate interesting and identify relevant trends in large groups of students who take the test, the same test results may be only a weak indicator of what any particular student knows or can do. The performance of all students in a school may point to areas in which curriculum and instruction may need to be improved, but an individual's poor performance may indicate an illness rather than his or her skill or knowledge level.

These observations lead inescapably to the conclusion: technologies are non-static. Because humans are always seeking improvements in technology (e.g. to meet emerging needs or to improve efficiency, safety, or reliability), all technologies are dynamic and constantly being changed. As new discoveries are made, the technologies that are used in one module or at one level of organization will be updated and reinvented. These changes will affect all of the technologies in connected modules and at all levels, thus the function of the entire technology system is affected by a single change. Technology also changes because of changes in social conditions. In education, the non-static nature of the technology is illustrated by curriculum and instruction which is changed to incorporate new expectations, new discoveries, different methods, different students, new skills developed by the teacher, or even the changing preferences of an influential educator.

Technologies tend to grow through accumulation. When a technology developed for one purpose is applied to another purpose, the receiving technology complex is said to be *structurally deeper* as it becomes more complex than it was before the new module was added. Structural deepening can arise from several different causes: Users of a technology will actively seek to improve the performance of the components of a technology; new technologies can be invented or existing technologies can be adapted to become new modules in an extant technology. Humans also discover new applications of existing technologies and find unintended uses of technologies so structural deepening can arise from exaptation.

The application of computers to curriculum and instruction is an example of structural deepening in education. Education existed, and it included its own collection of technology for instruction, prior to the invention of digital electronic computers in the middle of the 20th century. When the devices arrived in schools, education underwent structural deepening as new curriculum was added and as new instruction was designed to make use of the technology that had recently entered teachers' toolboxes. In addition, new layers of management were necessitated as network administrators and computer technicians were added to school staffs. Further, the need to support teachers as first they learned how to use and teach with the new tools deepened the professional development module of school systems.

Technologies are Non-neutral

Vannever Bush was a scholar involved with the invention and development of electronic digital computers. In his 1945 article "As We May Think," he predicted that computers would allow information workers to navigate and contribute to nearly infinite information pathways. He predicted workers would use a device called a memex to navigate and create paths through the mass of information. In many ways, the World Wide Web functions as Bush's memex; this is especially true of the web since the development and widespread adoption of web 2.0 tools and social media, through which users consume and create vast information.

The memex was perceived as a neutral technology. The adjective neutral captures the idea that the technology exerts no influence on the nature of the work completed with the machine. For Bush, consuming and creating information in the memex would be the same as consuming and creating information on print. The amount of information would change and using it would be more efficient, but the nature of the work and the nature of the workers' understandings would be unaffected.

The devices and networks we use do treat information as neutral. From a system design perspective, the goal is to move messages quickly, reliably, and securely regardless of the contents. The message "baby is a healthy girl" that comprises 22 characters including spaces is the same as "grandmother died today" when sent as text. Each requires the same computing resources to compose transmit and receive; we will ignore the predictive algorithms that can be used to compress messages. Those messages sent between siblings, however, would produce much different responses, so the information is decidedly non-neutral to humans.

Philosophers and scholars now observe that not only information, but also technologies are not neutral. The tools we have and how we use them influence how we act, interact, and—especially with information technology—think. Neil Postman, summarized the comprehensive influences of technology on human cognition: "New technologies alter the structure of our interests: the things we think *about*. They alter the nature of our symbols: the things we think *with*. And they alter the nature of community: the arena in which thoughts develop" (1993, 20) (emphasis in the original).

Technologies Transfer (Sometimes)

Most technologies are local creations. Members of a population will identify a need (or a discovery will illuminate a new need), they will find local naturally occurring phenomena that can be controlled and it will be applied to meet that need. Alternatively, they will exapt an existing technology to meet a need.

Different populations, even those with similar local resources may create similar technologies to meet similar needs or they may create different technologies to meet the need. Attempts to transfer a technology from one population to another are frequently met with difficulties that arise some several factors.

In general, technologies will transfer from one population to another when there is no preexisting technology for that domain in the receiving population but only if the culture is receptive to the changes induced by the arriving technology. Factors influencing the receptivity to a technology include existing skills and existing psychological or sociological taboos or preferences. Also, new technologies can pose threats to old expertise and existing political and social hierarchies that can inhibit the transfer of technology. Once a technology becomes embedded in a culture, attempts by authorities to prevent the diffusion of a technology through regulation generally fail.

Historian of technology, Arnold Pacey (1990) observed that, occasionally, technologies could be transferred from one population and region to another. The transfer is complete when the technology is adopted and the population adapts to its sociocultural influences. In other cases, the receiving population is slow to adapt to new technologies. In many cases, the arrival of new technologies (by transfer) is associated with a period of expanded innovation as the receiving population modifies the technology and invents new uses of the technology and new social expectations as a result of the technology.

The transfer of mobile phones into adult populations is an illustrative example of technology transfer. Many parents were motivated to begin carrying cell phones when it became clear that the only method of contacting their young adult children was going to be via text message or other messaging system provided by a smart phone.

Designing Solutions for the Wicked Technology

When designing solutions for tame problems, designers can apply systematic approaches. Problems can be deconstructed and solutions can be developed and tested in insolation before being introduced into the entire system. Further, the entire system can be tested before it is put into production. Because of the wicked nature of education, the methods used by designers of technologies are not appropriate for designers of solutions to education problems.

In reviewing practices that appeared to be most effective in designing solutions to wicked problems, Rittel and Webber (1973) recognized that different people perceive the problem (and its solution) differently, that experts sometimes have a too narrow view of the problem and potential solutions, and that wicked problem solving does not proceed in a linear manner. In this final section of the chapter, attention is given to the nature of solutions to wicked problems that tend to be perceived as sufficient by diverse groups within the effected populations. In general, planners who adopt these practices tend to create solutions to wicked problems that are judged sufficient by a greater portion of the population than planners who adopt the same methods used to design solutions to tame problems.

Recognizing Diversity

When Rittel and Webber (1973) first defined wicked problems, they concluded that leaders cannot reasonably identify a single solution to a given social problem. They observed "that diverse values are held by different groups of individuals—[so] what satisfies one may be abhorrent to another, [and] what comprise problem-solution for one is problem-generation for another," and in this situation, "there is no gain saying which group is right and which should have its end served" (Rittel and Webber 1973, 169). While factors that increase the diversity of the population served by planners complicates the work of designing solutions, imposing a solution or treating a heterogeneous population as homogeneous is not associated with the design of solutions as judged good by diverse subpopulations.

When a solution is defined such that one subpopulation finds it satisfactory but another finds it abhorrent; implementation leads to some being winners and others being losers. Game theorists call such situations zero-sum games. Many scholars who study wicked problem solving recommend planners attempt to design non-zero sum solutions to wicked problems. In these solutions, all individuals or populations perceive the solutions as advantageous. Those scholars recognize that the win may be disproportional for some, but this outcome is generally regarded as preferential to a zero-sum outcome. Those scholars also concur that the greater the number of choices and the greater diversity offered in the solution(s), the greater the non-zero sum potential of the solution(s).

Generalized Approach

Buchanan (1992) suggested that planners who are too specialized in their focus when solving a wicked problem are likely to limit the number and diversity of choices in the solution, thus decreasing the potential for non-zero sum outcomes. Buchanan also described an approach to designing solutions to wicked problems that integrates the expertise of specialists from many disciplines, but that make use of a new "liberal arts of technology cultures" (5) that bring more generalized skills and approaches to wicked solution design. According to Buchanan,

> Without integrative thinking disciplines of understanding, communication and action, there is little hope of sensibly extending knowledge beyond the library or laboratory in order to serve the purpose of enriching human life (6).

For Buchanan, experts' knowledge and understanding is essential to solving many wicked problems, but that expertise can prevent those individuals from being able to see the problem from other perspectives and that may prevent non-zero-sum solutions from being considered. The work of designing solutions for wicked problems proceeds from generalizations that can be applied to a wide variety of settings and it focuses on detailing how those generalizations can be instantiated using local resources to meet the particular needs of a local population (see figure 2.4).

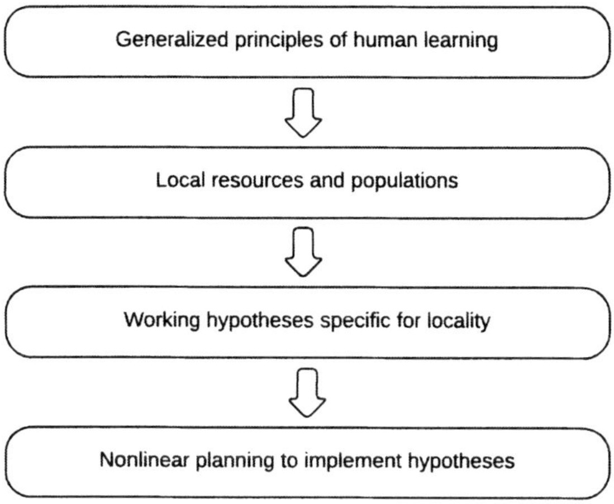

Figure 2.4. Generalizations lead to working hypotheses for local planning

As educators design solutions for teaching and learning, they begin with broad understandings of human learning such as the principles of deeper learning. In reconciling those ideas with the limits and opportunities of the local setting, the educator develops hypotheses predicting how the ideas can be instantiated in his or her classroom. As the plans are implemented, the educator begins an on-going and simultaneous assessment, design, and implementation; educators' expertise is in design and evaluations not in specifying solutions.

Non-Linear Planning

Tame problem solvers typically proceed in a linear fashion first articulating goals, then gathering data, engineering and implementing a solution, and finally evaluating the solution (see figure 2.5). Evaluation typically leads the planner to begin the linear process again in an iterative manner; what was learned in the one iteration is used to begin the next iteration. Curriculum and instructional planning that follows this model defines the goals to be met and then proceeds to develop instructional activities to meet those goals. Evaluation leads the educator to refine the plan for the next time the plan is implemented. Evaluation also leads the educator to decide if remediation is necessary, or if students are ready to proceed through the curriculum.

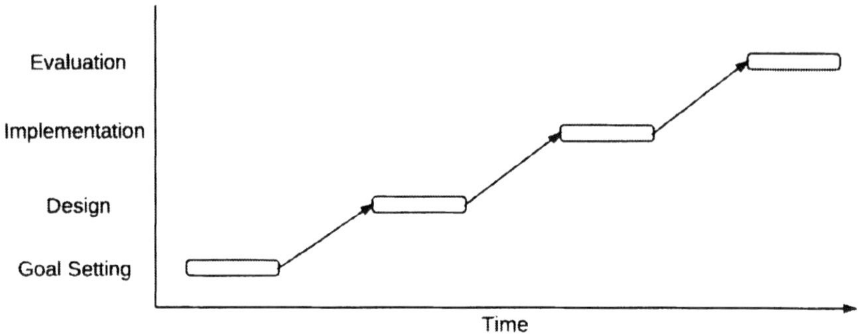

Figure 2.5 Linear planning for tame problem solving

Wicked problem solvers tend to engage in a different process. While the nature of the tasks is similar, the order in which the tasks are undertaken can appear random, and the boundaries between the tasks blur. The planning becomes an emergent process and it is modified to reflect the emerging circumstances of the implementation. Goals, designs, implementations, and evaluation methods are all dynamic in this method of planning and planners engage in each as deemed necessary (see figure 2.6). Far from being random, however, each step is typically supported by observations that can be clearly explained and can capture the conditions that led to the decision.

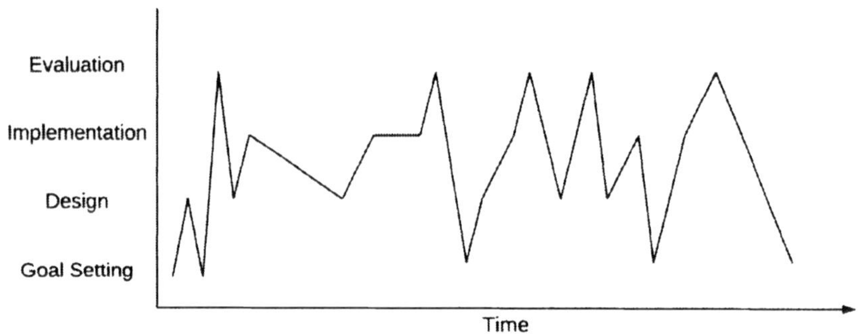

Figure 2.6. Nonlinear planning to solve wicked problems

Hermeneutic Approaches

In many ways, the seemingly erratic behavior of those designing solutions to wicked problems is similar to the approach used by hermeneutic researchers. Hermeneutics is a research tradition that originated with scholars who studied religious texts. These researchers sought to elucidate the meanings of sacred texts and they sought to understand the social context in which the texts were composed as a clue to the original meaning of the authors.

Anthropologists adapted the methods of hermeneutic researchers to understand cultures in terms of artifacts from the culture and to understand artifacts as a product of culture. Such methods are frequently applied to analysis of documents, tools, practices, and similar artifacts as data sources. Hermeneutic researchers follow the hermeneutic cycle (see figure 5.4) in which an artifact is interpreted in light of the culture and then the culture is reinterpreted in light of the emerging understanding of the artifact. This cycle between the whole of the culture being reconstructed and the parts of the culture embodied in the artifacts continues until the research can justify his or her conclusions.

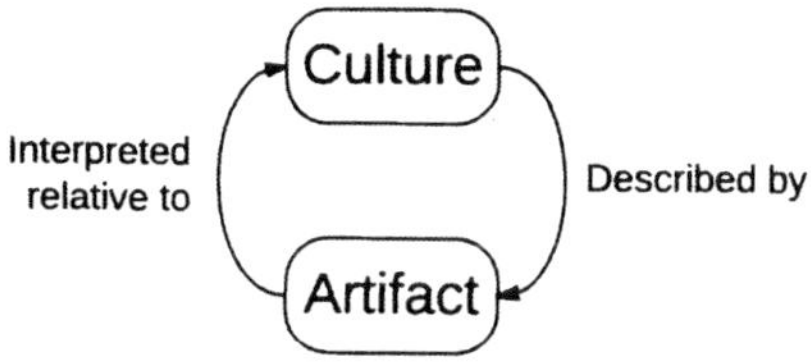

Figure 2.7. The hermeneutic cycle

Philosopher Shaun Gallagher (1992) articulates the connection between the hermeneutical approaches to understanding and solving problems related to teaching and learning:

> Things are not disjointed. Parts are parts of a whole. Learning does not consist of stumbling immediately upon an immediate, absolute, and satisfying knowledge of something. Learning is rather searching for understanding within a context (195).

From this, it is reasonable to conclude building understanding is an on-going and never-ending process of improving knowledge. As a human learns more, he or she better understands what is known, but also becomes more aware of what is unknown. The concepts of idea improvement will be encountered in the chapter addressing learning tasks as well as the chapter addressing improving system capacity. This approach is employed to develop ever-improving understanding of the parts and processes and goals of teaching as learning emerges.

Naturalistic Approaches

When planners recognize the wicked nature of their work and they adopt a design process aligned with that proposed by Buchanan (1992) and with the hermeneutic researchers, they will necessarily adopt a naturalistic stance towards the problems. In their seminal book on naturalistic inquiry, Yvonna Lincoln and Egon Guba (1985) argued that much scientific research is based on a reduction of the problem according to positivist principles, and that those assumptions are increasingly insufficient to describe many problems in the social sciences, including education. Whereas, positivist theory holds that a single reality exists and it can be understood through objective observation, naturalistic researchers accept an active role of the individual in creating and understanding knowledge and meaning. Whereas the positivist holds understanding of and control of causes can lead to predictable effects, the naturalistic researcher holds that in many situations causes are unpredictable and uncontrollable. Also, the positivist holds that observers can make independent observations of situations, but the naturalistic researcher is an active participant in the situation under study and the researcher is an active participant in data collection.

When adopting a naturalistic stance towards curriculum and instruction, an educator will recognize that many factors influence decisions and progress and that many sources of data can be used to demonstrate learning. In addition, these educators are open to and actively search for conclusions about learning that are supported by divergent data. Further, they seek alternative explanations for observation and apply inductive reasoning to identify and elucidate both know and unknown causes of observations. Naturalistic educators also recognize the active role that he or she plays in creating the learning environment.

Chapter 3
Learners

Leaners and their brains are the natural phenomena in which the technology of education is grounded. To be educative, an experience must be compatible with the physiology and psychology of their bodies and brains. For the 21st century educator, the classroom is filled with learners who have much different relationships with technology compared to those who entered even a few years ago, and this affects their physiology and psychology. The differences arise from the vast information and ubiquitous technology in which they are immersed (and have been since birth). Educators also understand the brains entering their classroom to a level not previously possible.

Human beings are unique creatures. We walk upright, and our freed forelimbs developed unusual dexterity allowing us to build and use tools. Because we walk upright, our pelvises are narrower than the pelvises of other primates. To accommodate birth through such a pelvis, human babies are born too small and helpless to support even the basic movements necessary for independent life until months after birth, and we are dependent on adults for years. During these years, we learn. We learn a lot.

We learn to recognize those who take care of us and we respond to their faces and voices moments after birth. We learn basic concepts of physics such as objects are solid, unsupported objects fall, and a moving object that strikes a still object will cause it to move. When crafty psychologists show us phenomena that appear to violate these physics, we react with surprise and stare longer at unusual situations than we stare at expected situations. We learn the basics of anatomy and are able to recognize (for example) our arms and we learn to move them and then control them at will. We learn to recognize emotions and we learn we can influence the behavior of others; "Dad smiles when I smile" and "Mom feeds me when I cry."

We learn the nuances of human interaction; we compete and cooperate, lie and cheat, share and support, we trade and beg and borrow and steal. We learn to share information; we ask for help and give help when asked. We express empathy and sympathy and disgust. We learn to sense others' states of mind and to communicate (and sometimes deceive others about) our state of mind. We act and react according to what we sense about the environment and what we learn about the environment from others. We learn to model our actions after the actions of others; from their examples we learn what to eat, how to avoid being eaten, how to protect ourselves from extremes of weather, to navigate the world, and to find and fill our role in society. We learn to capture the world in language and images, and we learn how to contemplate our place in the cosmos.

We learn to build physical tools to manipulate the external environment and to build conceptual models to explain and predict the environment (which includes our fellow humans), and these models become a very familiar (if unreliable) internal environment. We learn to test our models and modify them based on the results (or we explain away observations that are contrary to our models). Alone, we can begin to construct our models and conduct our contemplations, but we are even better at constructing and contemplating when we collaborate; much better.

Despite the central role of learning in human survival, the nature of learning is still an area of research that is very active. We are learning that behaviorism, the psychology in which instructionism is based, does not accurately predict and explain our observations of learners. This chapter focuses attention on the emerging understanding of human learning and on the influences of the information technology landscape on the learners who enter classrooms. Finally, the chapter describes the recent research regarding the digital generations whose members are deeply influenced by the technology they use.

THE NATURE OF LEARNING

Late in the 20th century, a diverse group of scholars (medical researchers, psychologists, computer scientists, philosophers, and others) started applying amazing new tools to the human brain. These tools include philosophical and epistemological tools (ideas to help us think about human learning), clinical and therapeutic tools (methods for studying patients in hospitals and similar setting), and laboratory tools (methods for conducting experiments and otherwise gathering data in controlled settings). Among the most interesting laboratory tools have been imaging tools such as computer aided tomography (CAT), electroenchalography, transcranial magnetic stimulation, and functional magnetic resonance imaging (fMRI) which have allowed researchers to study healthy and functioning brains under well-controlled conditions. Before such tools were available, much brain research took years to complete and was quite grisly as patients who suffered brain injuries were studied in great detail and then their brains were autopsied so that their disability could be associated with the part of the brain found to be damaged.

Modern tools are helping cognitive scientists understand how humans learn to a much greater level of accuracy and sophistication than has ever been known to humans, and the rate of discoveries in cognitive science is accelerating, so new discoveries displace current knowledge quickly. The speed at which these discoveries are being made is one of the factors complicating the work of refreshing education for the current and future generations. Among the important discoveries made by cognitive science are those that describe the importance of social interaction in learning, the physiology of learning, the plasticity in the human brain, and details of how the environment contributes to brain function.

Social Brains

Many species live in social groups and interaction within those groups is well known. In humans, however, social life takes on a level of complexity and sophistication that far exceeds what is observed in other species. Michael Gazzaniga, a noted neuroscientist who has studied human brains for decades, concluded "The shift to becoming *highly* social is what the human is all about," (emphasis in the original) and he continues, "Our higher intellectual skills arose as an adaptation to our newly evolved social needs" (2008, 111–2). He argued that, in natural environments where humans first lived, there was strong competition from other species for the resources needed by humans and there were strong predatory stresses on humans. In this environment, individual humans gained both survival and reproductive advantage by forming mutually supportive social groups. Humans' sociality affected the evolution of their brains. Mathew Lieberman, a psychologist from the University of California, Los Angeles, observed, "Our social nature is not an accident of having a larger brain. Rather, the value of increasing sociality is a major reasons for why we have a larder brain" (2013, 33). Having discovered the origins of humans' unusual brain development; cognitive scientists have turned attention to understanding the structures and functions of human brain.

Humans' social interactions depend on several complex brain functions including the perception of subtle signals, the recognition of patterns, the construction of and recollection of complex ideas, and the control of actions in response to all of these. Brains are compartmentalized organs; different sections are associated with different cognitive functions, and when a section of the brain is more active, several measurable changes occur in that section. Most social interaction is associated with increased activity in neocortex of primate brains. (The neocortex is the outermost layer of the cerebrum. You can locate the cerebrum under your skull by pulling a baseball cap down tight over your head—it will just about cover the part of your skull protecting your cerebrum.)

In primates, the size of the neocortex is positively correlated with social interactions; the bigger the neocortex, the greater social interaction observed in the species. Of the primates, humans have the largest neocortical regions. The social interactions that are associated with the size of the neocortex include the size of the group with which an individual can maintain grooming relationships, the ability of an individual to interact without using force, and the frequency of play. It turns out that "playing well with others" is important for human survival and human brains are designed to play well with others.

The highly social groups that characterize human populations depend on individuals' theory of mind. According to this theory, a human has the capacity to be aware of what he or she is thinking and at the same time understand that other humans are also thinking and the other may be thinking either the same idea or a different idea. Steven Pinker (1997), a professor of psychology at Harvard University, explains this ability to guess what others are thinking with these words:

> We mortals can't read others people's minds directly. But we make good guesses from what they say, what we read between the lines, what they show on their faces, and what best explains their behavior. It is our species' most remarkable talent (30).

As Pinker suggests, theory of mind is a component of the social brain appears to be uniquely human.

While theory of mind is an important aspect of sociality, it does not appear to be sufficiently sophisticated to capture the complexity of humans' social life. In his 2014 book, *A Natural History of Human Thinking*, Michael Tomasello elucidated the shared intentionality hypothesis, which comprises three elements. First, humans are capable of understanding a situation from different perspectives. Second, humans can self-reflect and can reflect on the intentions of others. Third, humans can evaluate outcomes against intentions and against social norms. It is not until all three of these characteristics developed that uniquely human cognition, which is embedded in culture, appeared.

Tomasello clarifies the role of social interaction in the development of an individual human as a learner as well. According to Tomasello, "a modern child raised on a desert island would not automatically construct fully human processes of thinking on its own" (2014, 6). While children are born with the capacity to communicate and collaborate and learning, Tomasello observes, "it is only in actually exercising these skills in social interactions with others (6)" that human cognition develops.

Some social scientists and educators understood the fact of the social brain long before natural scientists found evidence of it. Late in the 20th century, educators rediscovered the work of Lev Vygotsky, the Russian scholar who developed sociohistoric psychology in the early 20th century (see for example Vygotsky 1978; Moll 1992). According to this theory, knowledge is constructed in social interactions and influenced by the culture in which one lives. Learning, Vygotsky observed, is a cultural activity as much as a cognitive activity. Vygotskian thinking did influence many scholars, curriculum designers, and educators in the 20th century, but it has been abandoned in many instances as standards, standardized tests, and instructionism gained prominence in the first decade of the 21st century. Because human brains are organs designed by and for sociality, it is anticipated educational practices that reflect the realities of how brains learn will replace curriculum and instruction based on theories that ignore the social brain.

Brain Function

Like all organs, the human brain is made of tissue that comprises many different types of cells. The social interaction essential to human life is controlled by electrical and chemical activity in neurons, which are supported, protected, and nurtured by other types of cells. Neurons are elongated cells that become activated when an electrical signal propagates along the cell (figure 3.1). When the signal reaches the end of one cell, a chemical (called a neurotransmitter) is released from the axon of one neuron diffuses across the space, which is called a synapse, and it is taken in by the dendrite of the next neuron. The uptake of the neurotransmitter induces the second neuron to initiate an electric signal that propagates down its length.

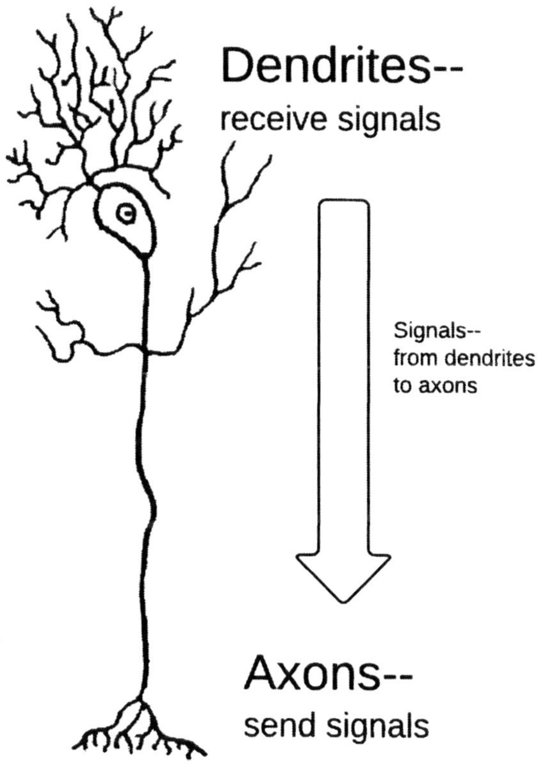

Figure 3.1. A typical neuron

Memories, skills, and many other varieties of knowledge are stored in these synapses; each time a signal proceeds across a particular synapse or along a neuronal path (when the synapses fire), a particular cognitive function or memory is active. When one learns how to (for example) solve a quadratic equation, the knowledge is stored in a pathway of neurons and synapses. When solving those problems again, the human uses the same neural pathway in his or her brain. This complex process has been simplified in the adage "what fires together wires together," and there is strong evidence that pathways are strengthened through repetition.

Brain Plasticity

Until recently, it was assumed that once a brain was mature (at the end of adolescence) it was an unchanging organ. Cognitive scientists have found that the human brain is actually quite plastic—cognitive functions that were assigned to one neural pathway can be reassigned to a different neural pathway. Early in life (through early adolescence) there is a period (lasting several years) during which many, many connections between neurons are established. This is followed by a period of pruning during which those connections that are not strengthened through repetition and practice are lost (Willis 2006). Later in life, pathways can be rewired, although the rewiring is usually more difficult than the wiring that occurs early in life.

The rewiring of one's brain is something with which I have some personal experience. I suffered a stroke when I was 42. Doctors were able to remove most of the clot from an artery in the base of my brain, but when they did, part of it broke off and entered "a very forgiving part of the brain." Those words were used by the brain surgeon to explain my condition to my family and were his description of a very plastic part of my brain. When I regained consciousness, my left side was quite disabled; I could not walk. With the help of some very talented therapists, I was able to regain the ability to walk, and my therapy became a lesson in brain function and plasticity.

My therapists gave me a long list of actions to follow when taking my first steps: "keep your toes up," "land heal first," "kick that leg forward," "head up," "back straight," "don't lean too far forward." It was hard work to keep all of those things in mind and I became thoroughly confused as I tried to perform each in the correct order during my first tentative steps. Walking, which had been managed by pathways that did not require conscious control, now required much effort and concentration (but still were slow, deliberate, and very clumsy). The pathways and synapses in my brain controlling the muscles for walking had been damaged.

For several weeks, my therapists had me perform what seemed quite silly exercises, (standing on one foot on a pillow with my eyes closed was a favorite), and they directed me to pay attention to how my body was reacting (for example when standing on a pillow with eyes closed, I attended to my ankles). I could sense my brain rewiring the pathways to control the muscles in my body, and by repeating the exercises I strengthened the connections. When my therapists took me to walk on gravel and grass, I could sense the same muscles and pathways working in my ankles that I had sensed when standing on a pillow. My experiences led me to the conclusion that the exercises alone would never have been sufficient for me to become a walker again; but without the exercises, my journey back to walking would have been longer and more frustrating. I have extended the lesson to education; we must make use of limited exercises and include many authentic experiences. Students must both stand on pillows and walk on grass.

The Environment and Brain Function

The brain functions that are associated with learning depend on communication with between neurons, which is mediated by neurotransmitters. Many chemicals are known to function as neurotransmitters, and also many chemicals are known to influence the functioning of neurotransmitters. Further, cognitive scientists have identified several environmental factors that are associated with the production and function of neurotransmitters and the neurons they connect. All of these factors influence how the brain perceives the world, cognates, and responds to the world; so they affect how humans interact socially and learn.

Several examples illustrate the effect neurotransmitters have on brain function. Glutamate is a neurotransmitter that crosses the synapses when two neurons first communicate, so it is essential in initial learning. Serotonin is associated with moods and emotions; abnormal levels of this neurotransmitter are associated with depression, anxiety, and similar conditions that can be debilitating. Drugs to control the level of serotonin reduce the symptoms of these conditions in many individuals. Norepinephrine is associated with levels of attention and motivation; too much or too little of this neurotransmitter can interfere with one's ability to pay attention to important parts of the environment. Dopamine is a neurotransmitter associated with one's perception of reward and with learning. Many addictions can be explained by the release of dopamine when the addict engages in the addictive behavior.

In addition to the role of many neurotransmitters that influence brain function, cognitive scientists have elucidated the role of structures in the brain that function in the control of humans' social interactions. The amgydala is associated with emotional responses to situations, and it "decides" if incoming information is important. Meaningful information is passed along to the hippocampus, which "decides" how to sort and store the information. In reality, all of the processes are far more complex, but even when oversimplified, educators have a more accurate and sophisticated understanding of human learning if they recognize that emotion and the environment affects brains.

Lieberman (2013) reported several findings that illustrate the modular organization of the brain and the adaptions of the brain for social interaction. When we engage in cognitive tasks, lateral (outer) sections of the cortex tend to be more active. When we engage in social activities, the medial (middle) sections become active. Scientists can increase lateral activity in the brain by asking a subject to perform a task; itself this is not very interesting. What is interesting is that when the brain stops the focusing on the task, it reverts to a pattern consistent with social interaction. It appears the default mode for human brains is to be socially active; when a brain is not otherwise engaged it is primed to attend to social interaction. Further, the same parts of the brain that are active when we feel physical pain are active when we feel social pain.

Several discoveries about the environment and brain function have particular relevance to educators designing exemplary practices for 21st century curriculum and instruction. First, moderate levels of exercise are good for brain function (Ratey 2008). Moderate is a difficult quantity to define, and the data are not yet conclusive, but it appears that the physician who recommends aerobic exercise that raises your heart rate to 80% of maximum for 20 minutes three times per week is recommending too little exercise. Fit bodies deliver food, water, and oxygen to brain cells and those are all necessary for proper function; fitness is also associated with greater levels of chemicals to build and support synapses. From this we can conclude that educational structures and organizations restricting students' opportunities to move around can interfere with healthy brain function, thus limit students ability to learn.

These findings support the inclusion of fitness-based physical education programs in the curriculum and these findings refute policies restricting participation in extracurricular athletics for academic reasons. Further, these findings suggest that decisions to reduce physical education classes to allow for enhanced academic instruction as a strategy for improving test performance are likely to result in the opposite effect. Students who are restricted from athletics are likely to have greater academic difficulties despite the increased time for academic activities.

Second, enriched environments are good for brains. This conclusion was first supported by research in which rodents were kept in two different environments; those kept alone and in cages without structures on which to climb or other toys had brains with fewer connections than were found in the brains of the mice kept together with other rodents in cages that contained complex toys and structures. This research has been criticized, as it was not clear the changes in the brains were the result of brains atrophying in stark conditions, brains in stark conditions failing to develop properly, or brains growing abnormally complex in the enriched environments. The implications for educators are identical regardless of the causes of the changes: enriched environments are associated with richer connections in brains and more connected brains are indicative of those that have learned more and that are capable of learning more.

Third, stress can be either good for brains or bad for brains, the effect depends on the level of stress that is experienced. What is true for physical stress (exercise) is true of psychological stress: none is bad, some is good, but too much is bad. Stress results in the release of a chemical called cortisol into the blood. Although not a neurotransmitter itself, cortisol does influence the release of the neurotransmitter norepinephrine, which causes the fight-or-flight response with its characteristic increase heart rate and more focused attention. In moderate levels, cortisol is associated with normal and healthy building of new connections between neurons and the brain functions necessary for social interaction. Too much cortisol can interfere with the normal building of connections and brain function. Exposure to excessive levels of cortisol for an extended time (for example when individuals experience chronic stress) is associated with degraded brain function.

We can conclude that modern cognitive science is creating a model of the human brain that is more accurate and more detailed than ever available to educators. This situation is not unlike the changes in medicine that followed the discovery and elucidation of germ theory. Once scientists discovered germs and understood their role in disease, doctors began to take steps to minimize the spread of germs; for example, they started washing their hands frequently. It is anticipated that educators who ignore the effects of their classroom environments on brains will be viewed with as much suspicion as doctors who ignore the advice to wash regularly.

Brain-Friendly Classrooms

In addition to the physical and psychological conditions associated with their classrooms, educators must recognize the conditions of their classrooms that are associated with increased attention and learning. One of the most meaningful discoveries of cognitive science for educators is the importance of motivation and emotions in attention and cognition. The relationship can be oversimplified (but with accuracy) with the statement included in John Medina's (2009) *Brain Rules*, "we don't pay attention to boring things," and we cannot learn if we do not pay attention.

Several research groups have investigated the characteristics of classrooms associated with increased interest. Through this research, scholars are finding positive association between motivation, cognitive engagement, and learning. Blumenfeld, Kempler, and Krajcik (2006) summarized the aspects of curriculum associated with motivation and cognitive engagement:

- value- Learners value curriculum when they enjoy the task, when they see clear connections to the their lives outside schools, or students otherwise find the task relevant.
- competence- Learners who view themselves as able to perform the task and who have greater levels of self-efficacy demonstrate greater levels of competence.
- relatedness- Learners who feel they belong to a community whose members are concerned with his or her well being have greater levels of relatedness.
- autonomy- Learners who have the opportunity to influence the selection of curriculum and methods have greater levels of autonomy.

Blumenfeld, Kempler, and Krajcik note, as well, the factors associated with motivation and cognitive engagement are both self-reinforcing and positively associated with the others. When one factor increases, the others tend to increase as well producing a positive feedback loop.

One of the challenges for educators seeking to create brain-friendly classrooms is to ensure that individuals are coached and supported through his or her zone of proximal development. Lev Vygotsky defined the zone of proximal development (ZPD) as a psychological idea, and educators have used it as a construct to understand what comprises appropriate curriculum. The ZPD can be illustrated on the graph shown in figure 3.2. On this graph, skills are plotted on the y-axis from simple to complex. Time is plotted on the x-axis. Two parallel lines with positive slope represent the ZPD, and divide the skills into those "above" the ZPD, those "below" the ZPD, and those within the ZPD.

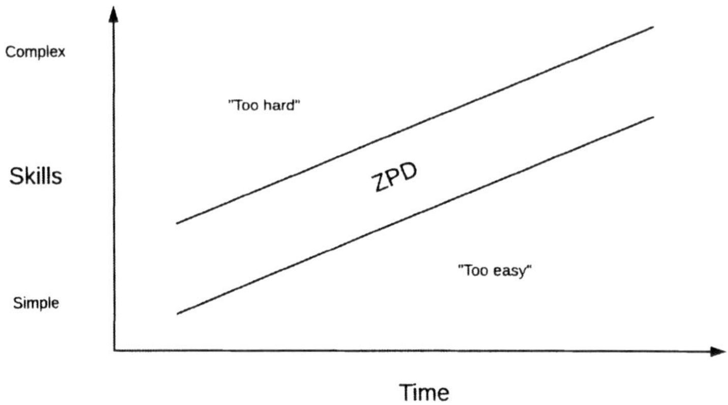

Figure 3.2. Vygotsky's zone of proximal development

Those tasks that fall below the ZPD are interpreted as being comfortably within the abilities of an individual. Learners understand those parts of the curriculum and can complete problems with independence. Those tasks that are above the ZPD are incomprehensible to the learner. The positive slope of the ZPD indicates that over time and with greater access to educative experiences included in the curriculum, students are capable of independently completing more complex tasks the curriculum. Central to the concept of ZPD is that only those tasks that fall within the zone can lead to greater learning. Those below the ZPD are too easy; those above are too hard. Learning only occurs within the zone.

The tasks that fall within the ZPD are generally understood to be those that can be completed by the learner with social assistance. Feedback, coaching, and scaffolding from mentors such as teachers, discourse with peers, and the interior dialogue associated with reflection have all been effective in providing the social interaction associated with successful completion of tasks in the ZPD.

HUMANS AS TECHNOLOGY-USERS

Technology is a relatively new word to the lexicon. Jacob Bigelow, a New England botanist and doctor who published a series of his lectures as a textbook, first used the word in print in 1831. Evidence of technology use by humans, however, extends far into pre-history. Wherever archaeologists find evidence of ancient humans, they find evidence of the technology used by those humans, and scholars know some ancient human cultures primarily by the evidence of their technology. Given the observations of Timothy Taylor and Kevin Kelly on the inseparability of humans and their technologies (see pp. 26–27), we would expect deep understanding of technology. Despite this, the role of technology in human history is still open to interpretation and the effects of information technology on individuals and groups and cultures continue to hold the attention of scholars.

Technology and Brains

Paradigm mediums are aspects of society and culture that are so deeply embedded in culture that it is difficult for individuals and groups to perceive the influence of each on information and interaction in the society. Each paradigm medium serves an important function within society, but each also has both unintended and unrealized consequences. Commenting on them, Brad Mehlenbacher, a scholar associated with several programs at North Carolina State University, observed,

> Because paradigm mediums . . . form the very core of our systems for understanding, conceptualizing, and promulgating knowledge about, with, and into the world around us, they are exceedingly difficult to understand, isolate, parametrize, or control (2010, 7).

Information technology is a paradigm medium that has influenced humans' cognition in unrealized ways (Feenberg 1999). Until humans invented writing; money, sacred texts and thus religion, unchanging laws, and property were unknown (Goody 1987). These social structures and organizations arose only after writing provided more permanent memory. In societies where it is available, writing exerts an active influence on what individuals perceive to be natural aspects of human life.

Despite evidence that brains remain plastic into adulthood, human brains do adapt to the patterns of cognition that are experienced when young, and those become the basis of what individuals expect from other people. The technologies that one uses during his or her adolescent years and the habits associated with them form the basis of what he or she perceives to be natural and all other technologies are perceived to be artificial. Looking back, individuals tend to judge earlier technologies as primitive and a step contributing to progress that culminates in the natural technologies. Looking forward, individuals tend to judge emerging technologies as artificial and their use to be superfluous (see figure 3.3). The use of emerging technologies is usually judged as degrading human cognition, especially because cognition is transferred to technology. In reality, new cognitive skills replace old skills as new technologies emerge, and in many cases the new skills are more sophisticated than those previously necessary.

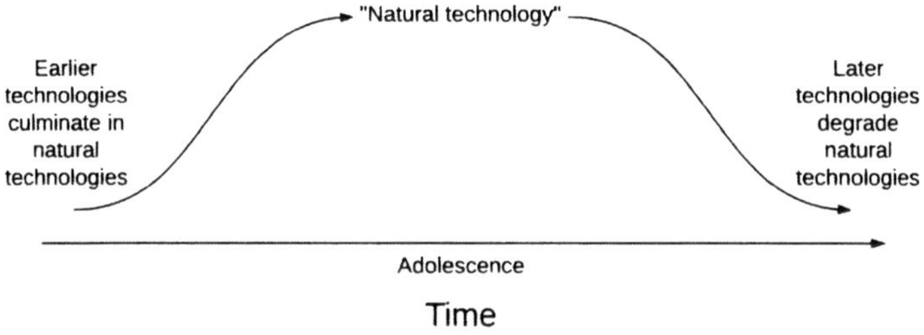

Figure 3.3. Technologies experienced during adolescents are perceived as "natural"

Bruce Wexler (2008), a psychiatrist from Yale University, suggested the experiences from one's interaction with natural technology are used to construct an internal cognitive structure of skills, knowledge, habits, and attitudes as one matures through adolescence. Because age cohorts experience approximately the same sociocultural milieu through adolescence, perceptions of and expectations resulting from natural technologies tend to follow generational patterns. When a human finds himself or herself in a setting that is contrary to what is natural, he or she feels dissonance and the setting (or some aspects of the setting) is perceived to be artificial. This explains why many adults decry (for example) the decision to introduce calculators into math instruction. The adults are so familiar with the technologies they were taught for doing mathematics that they do not perceive it to be a technology, which serves the same purpose as the calculator. The perception is that the calculator is an artificial technology and allowing students to use it will result in a degradation of essential skill.

With the transition from a print-dominated to an electronic digital technology-dominated society, scholars are beginning to rediscover that information technology influences how we frame, understand, and solve problems. Science and technology writer James Gleick (2011) recounts several stories in which business and industry leaders dismissed the commercial and social potential of telephones as they emerged and then eventually displaced deeply entrenched methods of communicating. This was in large part due to the deeply embedded business practices that existed before the telephone. Those practices created the communication paradigm medium of business and it was inconceivable to many that those natural practices would be replaced. It will become clear that the 21st century educational practices cannot ignore the influences of information technology on students. Students' experiences with technology affect how they think and how they interact, and so the cognitive effects and social expectations of technology-rich culture intrude into each classroom as students enter.

Technology Changes the Brain

Gary Small, a cognitive scientist from the University of California Los Angeles and his coauthor Gigi Vorgan summarized many studies in their 2008 book *iBrain: Surviving the Technological Modification of the Modern Mind.* The collection included several that documented the effects of environment on human brains. They described research in which scientists measured a larger portion of the brain controlling the right hand in expert violin players compared to other expert musicians and research that observed London taxi cab drivers have a greater part of their brain dedicated to controlling spatial visualization than control subjects.

Small and Vorgan even summarize Small's own research in which he created images of subjects' brains while they were performing computer-based information tasks. The researchers compared the images of experienced computer users' brains to images of those who were inexperienced users of computers. In the images from the tech-savvy group, the dorsallateral prefrontal cortex was more active than other parts of the brain. In the non-tech savvy individuals, the dorsallateral prefrontal cortex was not active beyond what was observed in other parts of the brain. After the initial images were made, the non-tech savvy individuals were exposed to five hours of training and practice in using computer-based information. Both groups then performed the original task again, and the brain images of both groups showed elevated activity in the dorsallateral prefrontal cortex. This suggests that very little exposure to digital media can cause a measurable change in how the brain functions.

Another phenomenon arising from exposure to network-connected digital devices that influences human brains is continuous partial attention (CPA). This term is used to describe the condition that arises when an individual is always alert for the beeps and buzzes that notify carriers of mobile device that a new message has arrived. The problem can be particularly acute for those who carry a single device that connects to multiple messaging systems. These devices are commonly found in the pockets of students in public schools, and CPA arising from those devices can interfere with attention to instructional tasks. Continuously sensing the vibrations (or similar signals of arriving messages) can draw individuals' attention from other tasks and contribute to cortisol-releasing stress. CPA has been associated with a decreased sense of control and with decreased levels of self-esteem (Ratey 2008). These effects have been associated with changes in the way the hippocampus (the part of the mid-brain associated with memory formation) functions.

On occasion (actually with startling frequency), I encounter educators who are conservative in that they want students to experience what they experienced. These individuals tend to share the idea that modern technology is degrading the cognitive skills of young people. As a result, they insist their classrooms become technology-free. While well intentioned, these educators are ignoring the fact that technology is affecting their students' brains and technology creeps into their classroom with their students. The members of the digital generations know only a culture enriched by digital technologies, and their brains are adapted to those environments.

21ST CENTURY LEARNERS

It is clear that the second generation of digital learners will soon be enrolled in schools. Most researchers mark 1990 as the year when the first digital generation was born. Children born in that year first enrolled in school when the Internet was arriving in schools and they were becoming adolescents as Web 2.0 with its interactivity was emerging. Some simple math will show that early in the second decade of the 21st century, the children of those born in 1990 (who will be members of the second digital generation) will be entering schools, and it is anticipated that the characteristics of the first digital generation will be even more deeply embedded in the second generation.

All human education systems, both formal and informal, reflect the culture that learners are being prepared to enter. Currently, we are observing one culture replacing another, as print-dominated information is being replaced by electronic digital information. Whereas printed information requires reading and writing, which are specialized skills usually developed through formal teaching, learning to use digital technologies can be effectively and efficiently learned through informal methods. These technology-rich informal learning experiences are influencing learners' expectations of classrooms.

Social scientists have developed a collection of tools for collecting and analyzing data about societies, including those for studying cultures and groups other than the one to which the researcher belongs. Commonly called ethnographers, these scientists recognize that the tools used to measure a factor in one social group may be useless in another (the tools may measure something different or a tool may be measuring a factor that exists in the researcher's culture but not in the culture under study). By using ethnographic tools, the researcher attempts to gather information about a society and culture and then interpret data free from bias introduced by the researcher's native culture.

In *Hanging Out, Messing Around, and Geeking Out*, a 2010 report that emerged from The John D. and Catherine T. MacArthur Foundation Series on Digital Media and Learning, Mizuko Ito and colleagues described the work of the Digital Youth Project, which was a collaborative effort to apply ethnographic methods to investigate the manner in which young people are influenced by technology. By deciding to use ethnographic methods the research group recognized the existing instruments for collecting data may not be valid for populations of the digital generations or that existing tools may miss important but unknown factors related to the first digital generation's interactions with digital devices. In this section, we consider several characteristics of digital generations that differentiate them from previous generations.

The Skills Inversion

For much of the 20th century, educators were adults who had earned an undergraduate degree, which typically requires four years of study in higher education, to become qualified. As undergraduate students, these adults had become skilled users of print information. As a result, educators were the most skilled users of the dominant information technology in classrooms. This is not the case in the 21st century.

Combined with young people's quick acceptance of new technologies and their tendency to multitask, it is likely the students in a typical 21st century classroom are more skilled using the networked devices and the online information sources than their teachers. Some educators perceive this as an opportunity to learn and ask students for tips and tutorials; some educators see this as a threat to their authority and will avoid digital tools as a result. This skill inversion (see figure 3.4) does threaten to disrupt much of the school organization that became so common in the 20th century.

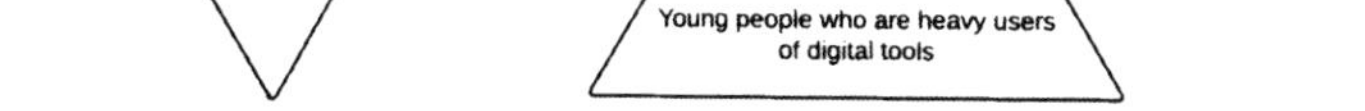

Figure 3.4. The skills inversion

For several generations, young people have been heavy users of media. Radio and television programmers and marketers paid attention to the young people (and they continue to), and generations of youngsters have been told by parents and teachers to "turn the television off" when they are doing homework.

In the 21st century, however, young people have begun using technology in ways that previous generations did not. Digital generations consume and create more (far more) content than previous generations, they interact widely through online spaces, and these experiences influence how they interact and how they learn in school and out of school. Ito and her research group used the terms "hanging out" to capture the central role of friendship in young people's use of information technology. They use "messing around" to capture the importance of interest in motivating their participation; and they document how that can lead to "geeking out" in which young people develop great expertise in their areas of interest.

Computers Becomes Mainstream

Just as young people were heavy users of analog electronic media (e.g. radio, television, recorded music), they also started using computers when they entered the consumer and education markets, but the highest levels of use were concentrated in rich white male populations. The stereotypical "computer geek" was based on the demographic profile of the people who were heavy computer users. This can, in part, be explained by the observation that the first computers to enter the consumer and education markets were designed for users to write programs. Landauer (1997) noted that programming requires certain cognitive skills, so it is especially appealing to those who are more introverted and who engage in more intuitive thinking than the general population. Early computer users tended to interact with friends and family less than their peers, and they reported less participation in extracurricular activities at school and less social activity out of school than the rest of the youth population.

In 2005, researchers supported by the Kaiser Family Foundation (Roberts, Foehr, and Rideout 2005) reported that young people who reported heavy use of computer also reported spending time with friends and family, participating in extracurricular activities, holding jobs, and engaging in physical exercise at a level never observed before. The researchers interpreted these results as computers becoming more widely used among young people; "computer geeks" were not necessarily participating in more activities (although there is some evidence they are), the young people who had previously avoided technology were actively adopting the devices in great numbers.

The widespread acceptance of computers by the young people can be explained in part by the increased usability of the devices that was observed over several decades. Originally, computer input and output was limited to alphanumeric characters including punctuation. By the early 1990's, alternative input devices such as a mouse that was used to control icons representing files displayed on a graphic user interface (GUI) and programs written by others to perform useful functions were widely available. By the end of the 1990's computers capable of displaying media (high resolution graphics, audio, video, and animations) and connecting to networks were widely available. Each advance in computer capacity increased the appeal to wider audiences.

The current media landscape for young people includes very mobile devices. Cell phones and handheld computers that include the capacity to connect to wireless networks are gaining popularity and are influencing students' experiences and expectations as information users. These devices also provide an excellent example of the phenomenon that science and science fiction writer Stanley Schmidt (2008) called "the digital melting pot" as mobile devices provide several services that were only available in separate devices to previous generations. Apple's iPhone provides an illustrative example of the convergence of phone, camera, notebook, book, and email device into a single handheld (clearly this is an incomplete list of iPhone functions and any list is necessarily incomplete as the collection of applications users install to customize their phones is growing constantly). As computers evolved from being a tool for programmers to multi-function mobile devices, they became tools for rich social interactions and thus became appealing to and used by very broad populations.

Multitasking

The same researchers supported by the Kaiser Family Foundation who found computers were becoming mainstream documented an increase in young people's increasing consumption of media by consuming more than one stream at the same time. They used the term Generation M to describe the young people who were consuming amazing amounts of media in reports published in 1999, 2005, and 2010. In the preface to the 2010 report, the authors commented on the 2005 report, and observed, "At that point, it seemed that young people's lives were filled to the bursting point with media" (Rideout, Foehr, and Roberts 2010, *x*). In the next sentence, however, they previewed the data that had been collected five years later, with the comment, "Today, however, those levels have been shattered" (*x*). The authors proceed to give evidence that young people in their sample were consuming media for seven hours and 38 minutes per day and through multitasking, they consumed 10 hours and 45 minutes worth of media each day—seven days per week.

Multitasking is the term used to describe the situation in which an individual is consuming and using more than one media at a time; the youngster who is reading a textbook (one medium) while listening to the radio (two media) and downloading music from the Internet (three media) is multitasking. (In reality, a multitasking teenager is likely to be engaged in several additional media tasks including those requiring interaction.) The phenomenon has been documented in many studies and the effects of multitasking on people who engage in the patterns of media have been the focus of much research, but the results are equivocal.

Psychologists and others who study multitasking and its effects on human attention, learning, and cognition have yet to decide if the effects are "good" or "bad." Much of the difficulty comes from the differences between the observations made in the highly controlled environments of the laboratory and the observations that are made in the real world (in natural settings). In the laboratory, subjects who are multitasking are measurably (and significantly) slower at almost any cognitive task that researchers devise. Scientists attribute the slowdown to the time necessary to activate different neural pathways to switch between the tasks. In the real world, the time limit imposed in the laboratory disappears, so the additional time necessary to complete a task does not appear to interfere with a multitasking subject's performance of cognitive tasks. In real world settings, multitaskers preform as well as others, but they take longer to complete the tasks. (This finding is applied only to tasks designed to be similar to academic tasks. Multitasking can be dangerous when one is driving an automobile or performing similar tasks that require focused attention and quick response. In those situations, the conclusion that multitasking individuals are "as good as others, but slower" is meaningless.)

There is evidence that this trend is extending and expanding; more people are multitasking with more devices in more places. There is evidence that multitasking is the preferred method of engagement for many young people; they like the social connections and their brains have adapted to multiple sources of information, but there is evidence that multitasking can interfere with attention (Kraushaar and Novak 2010; Quan-Haase 2011).

There is evidence that adults are less facile than young people at multitasking and schools are taking steps to minimize students' opportunities to multitask. Rosen (2010) suggested this may not be the appropriate response for educators, however, and concludes,

> The bottom line is that our students are multitasking and we cannot stop them without placing them in a boring, unmotivating environment. The trick is to develop educational models that allow for appropriate multitasking that improves learning (95).

The reasoning supporting Rosen's conclusion is simple and appears to be supported by evidence from cognitive science: Young people and their brains are immersed in an environment of multiple simultaneous information sources, so their brains become adapted to that stimulation, and they perceive such information and interaction rich environments as natural. Researchers appear to be converging on the conclusion that curriculum and instruction can be adapted to students' pattern of multitasking in a manner that promotes academic learning through curriculum-related multitasking. Classrooms in which educators adopt computer technology and engage students with curriculum and instruction that reflects the information and interaction-rich environments in which they live will be creating environments the members of the digital generation perceive to be filled with natural technology.

New Norms

Cultural activity is understood to influence cognition and expectations at various levels within the culture: from the individual to small groups to large groups.

The sociocultural context arises in large part from the milieu of technologies (especially information technologies), practices, and traditions that form the cultural environment in which humans live (Nasir and Hand 2005). This defines the experiences that become the expectations of learners who enter classrooms. There are several norms that are emerging in the digital generations and that have been associated with these generations' experiences with information technology, and educators will find it necessary to reconcile these aspects of the digital generations with exemplary classroom practices. With the arrival of computers, scholars and the audiences for whom they wrote also observed the beginnings of new trends in media use among young people (and older people as well).

Creators

Prior to the arrival of computers, young people had primarily been consumers of media. Once computers arrived, young people began writing and sharing programs, thus creating and disseminating digital media on a wide scale for the first time.

The digital generations appear to have a still-evolving relationship with media they create and consume, and within their patterns of use we see contradictions that include ethical aspects. Don Tapscott (2009) identified integrity as one of eight characteristics of digital generations, and he suggested that individuals in the first digital generation are concerned with "being honest, considerate, transparent, and obligated by their commitments" (82). Also, he cited evidence that the first digital generation is tolerant of differences and deeply committed to social, environmental, and similar causes they believe are for the general good. Despite the concern for the public good, this generation is also a group that uses their media skills to copy and share music and other media in violation of traditional copyright laws. They appear to have adopted John Perry Barlow's (1996) ethic of free information and free speech on the Internet, and many are willing to share their intellectual creations through Creative Commons and similar copyrights.

Virtual is Real

Mehlenbacher (2010) observed, "transformative technologies take their most dramatic shape when the general population interacts with them" (11). This observation is well illustrated by the evolution of the Internet. The Internet was first designed as a tool for academic purposes and the audience was limited; once the World Wide Web was established the Internet became a tool for the general population the Internet began to transform society. After web 2.0 technologies (such as instant messaging, blogs, discussion forums and similar tools) evolved and facilitated content creation and sharing along with social interaction by all users, the Internet contributed to revolutions in commerce and culture that are not yet complete (and that seem in a state of perpetual renewal).

Many scholars have observed and commented on the growing phenomenon of online presence that became common as users became content creators. Adults in the pre-digital generations are generally reluctant to create and maintain an online presence, and they tend to be very cautious about what they post. This is particularly true of educators. Members of the digital generations who grew up in a world rich with personal computers, laptops, the Internet, and mobile computing devices are quick to create online presence and they continue to maintain an online identity, and demonstrate much less concern about the contents of their online presence than the older generations. They establish and nurture rich social connections in their online lives; for them their virtual friends are their real friends.

Generally, young people maintain their online presence via social networks. Facebook is the most familiar; although it is difficult to verify the numbers of individuals who have Facebook pages, the entry for Facebook on *Wikipedia*, indicated there were about 864 million members late in 2014. Although the terms of service of Facebook and similar sites restrict membership to those over 13 years of age, there is no reliable method of ensuring that those who maintain sites are of the correct age. Adults find many of the private details that are part of the digital generations' online presence to be distasteful, disadvantageous, or detrimental. In some cases youngsters' online presence has contributed to them being in dangerous and even deadly situations. Despite this, an online presence is an important part of the digital generation's social life and their social identity, so it is anticipated that this will be a part of students' formative experiences long into the future.

Early in human history, an individual's identity was created by and for the people with whom the individual lived, and this number was small. Anthologist Robin Dunbar (1992) concluded that the size of the human neocortex limits the size of humans groups that can be maintained to about 150 individuals. Creating a new identity was possible if an individual left one group and found another to join; anthropologists have documented how marriage practices in many cultures encourage this type of movement between groups for young adults.

With the arrival of writing, laws and other permanent records became possible especially in populations that abandoned nomadic for agricultural lifestyles. These changes led to the establishment of individuals' legal identities that are more permanent than those found in cultures that have no writing. It was possible still for each individual to maintain multiple identities simultaneously, however. As a teen becoming an adult late 20th century, it was easy for me to create and maintain separate identities: My identity as a high school student was separate form my identity as a family member, and those were largely lost as I became a college student, and then an adult with an adult sibling. Further, my family identity changed, as I became a spouse and then a parent. I also had a professional identity that was separate from my other identities. Those identities did occasionally merge. When I worked in schools that enrolled youngsters whose aunts, uncles, and then mothers and fathers, had known me in high school; I had to explain the story behind my high school nickname. On those occasions, my long-forgotten identity as a high school student was revived and had to be reconciled with my professional identity.

Interestingly, in the digital world, it has become possible to maintain many different identities as well; these identities can be imaginative and even contrary to any physical identity. There are thousands of online communities that focus on any imaginable topic. Joining those communities (usually) requires only an email address, which can be obtained with relative anonymity from many sources, and it is very difficult to verify the physical identity of anyone who is a member of an online community. Sherry Turkle, a sociologist from MIT studied computer users' sense of identity early in the days of Internet-mediated communication. She observed that many users at were creating multiple online identities and exploring different senses of identity through online spaces. Turkle began her book *Life on the Screen* with the observation,

> At one level, the computer is a tool. It helps us write, keep track of our accounts, and communicate with others. Beyond this, computers offer us both new models of mind and a new medium in which to project our ideas and fantasies. Most recently, the computer has become even more than tool and mirror. We are able to step through the looking glass. We are living in virtual worlds. We may find ourselves alone as we navigate virtual oceans, unravel virtual mysteries, and engineer virtual skyscrapers. Increasingly, when we step through the looking glass, other people are there as well (1995, 9).

Turkle's observations proved to be an accurate prediction about the future of the Internet as it was an observation of the initial days of the Internet. Today, members of the digital generations are creating an online presence at an early age, and they demonstrate an openness and comfort in living online in a manner that is disconcerting and perceived to be unnatural to those who belong to pre-digital generations.

The dynamics of online communities has extended to the point where each individual's online presence is created and maintained by the online community itself and it grows with neither the knowledge of, nor the consent of, the individual. John Palfrey and Urs Gasser (2008), scholars from the Berkman Center for Internet and Society at Harvard University, introduced the term digital dossier to refer to the collection of digital records kept about an individual. For many individuals, a digital dossier begins when a pregnant mom posts the image from a sonogram on a social network site, and it grows from there. It also grows because users "tag" files with names of people (or other relevant information), so an individual's name can be associated with an image or other content that was posted without the subjects' consent.

There are other systems whereby information is collected by technology and about the individual using the technology. In the spring of 2011, several bloggers complained that the location of their cell phones was logged in a file on the phone (Allen and Warden 2011). With that log, a record of the travels of the person carrying it was being kept. Also, many retailers use radio frequency identification devices for inventory control, but there is potential to track items even after consumers have left the store. Ostensibly, this information is not being collected about an individual, but many find the potential for invasion of privacy via information about the final location of purchased items to be disconcerting.

Control

Early in the history of electronic digital computing, reprogramming a computer required technicians to physically reconfigure the circuits following the direction of the computer engineers. Lohr (2001) observed the insight that both the data and the directions for processing the data could be loaded into the same circuits had a profound effect on computing in the years after World War II. That advance led to the general computing devices that made their ways onto desks and ultimately into pockets.

With the arrival of the operating systems for personal computers that included a graphic user interface (GUI) in the 1990's, the metaphor of the computer screen as a desktop became common, and that term is even used to describe the default screen that loads when a personal computer boots up and is ready for use. Just as workers customize the physical desk at which they work with tools that help them accomplish tasks along with meaningful pictures and doodles and avatars representing themselves, those same workers customize their computer desktops. The desire to take this step to customize computers appears to be very deeply embedded in the computer-using culture as first and second graders are frequently distracted from projects they find engaging to change the background of their desktops. Their classmates hold those individuals who can perform the change in high esteem, and an operating system update that replaces favorite desktop images is not well received by those students.

As computers with GUI operating systems entered the consumer market, the number of software titles expanded as well. Consumers can buy a computer and then install software so that it performs the tasks they want, and also customize the interface so that it both matches their personality and is easy to use. Customization of a single computer is possible through user profiles—each user has a unique profile with its own customizations. On some networks, one's user profile is stored on the network, so a user's desktop is available on any workstation connected to the network.

The ability to customize a device to meet the user's interests and needs has continued to be a feature as computers moved into users' pockets. Since introducing the iPod in 2001, Apple Computer has been an industry leader in portable computing. With the iPod Touch and the iPhone, Apple—along with all of the other manufacturers of handheld personal digital assistants and computers—has continued the trend of customization of technology that can be traced to GUI-based desktops and software. Users of these devices (and anyone exposed to advertisements for the devices) quickly become familiar with the concept of the "app," which is a shortened version of application. Apps add functions to the handheld and are highly designed to perform a particular function, but usually allow the user to customize the information and how it is delivered. For example, an app for accessing information regarding the weather can be configured to show local weather or the conditions at any other location the user specifies.

Don Tapscott (2009) also observed that the ability to customize technologies has become deeply embedded in the digital generations, and he points to the customization of media they consume as an example of this character of digital generations. Whereas previous generations consumed the media that was provided by publishers or broadcasters, members of the digital generations create playlists and use similar strategies to access and consume exactly the media they want, when they want, and where they want. Tapscott also suggests this trend can be seen in non-information technology industry as members of the digital generation (only the first one, the second generation is not yet old enough) are a significant part of the customized automobile market, and that some brands of automobiles have emerged and are being marketed to this generation.

Speed and Innovation

Related to the digital generations' interest in and desire to customize technologies is the rate at which the digital generations adopt new technologies. Members of these generations are willing to buy new devices as soon as they arrive on the market and they are enthusiastic consumers of innovative new devices. They both become users of new technology quickly and they find new uses of technologies quickly.

As I was drafting this paragraph, I observed a situation in my classroom that illustrates the generational differences in the rate at which innovative technologies are adopted. I happened to be introducing, to a group of students and a group of teachers, an online tool for creating presentations; for both groups I introduced the session as "an interesting alternative to PowerPoint." Each group had similar introductions to the same system (a 10-minute session watching me use the tools on a projected screen and then about 30 minutes to work with the system themselves). Whereas the teachers expressed an interest in a more detailed training session and a discussion of how this was more effective than what they were already using (the precautionary principle manifest), the students began using the tools immediately (most stopped watching my demonstration well before I had finished and started their own explorations). By the end of the week (the introductions were given to both groups on the same Tuesday), several students had used the new system to complete assignments for other classes (some of which were taught by the same teachers who had been in the training session), but none of the teachers had even logged on to the system for a second time.

Several research groups have concluded members of the digital generations expect responses to communication far quicker than members of previous generations. Whereas previous generations ordered products from mail order retailers and expected to wait several weeks as their order was delivered to the retailer via postal service, was processed and filled, and then shipped back; digital generations expect to place an order on a web site and receive the product within a day or two. Whereas previous generations expected to take a test in class and have it returned graded within a day or two, the digital generations expect to take a test online and see the results immediately. Whereas previous generations called friends on the telephone and expected them to call back within a few hours if they were not available, members of the digital generations expect their friends to carry cell phones and answer their messages (phone calls or text messages) within seconds.

Implications for Education

Students enter school with very diverse backgrounds and social expectations that arise from their cultural backgrounds, and global mobility is increasing the cultural diversity of many communities. Students also enter classrooms having experiences within the sociocultural milieu that includes rich experience and expectations regarding networked digital technology and information. This contributes to the growing recognition that the learners in K-12 classrooms in the 21st century are not the same learners that were in the 20th century classrooms. Educators will recognize and leverage these aspects of young people as communicators and learners as they designing learning environments.

In Western cultures, the history of education is marked by a differentiation between everyday experiences and the explanations and language used in schools. Lev Vygotsky observed and explained these differences. Everyday learning occurs primarily through speech and proceeds from informal sensory experience to formalized generalization. Schooled learning occurs primarily through written language and proceeds from generalization to sensory experience. By connecting everyday experience to learning in schools and by using information technology tools that students perceive to be natural, educators can make school meaningful and can lead to deeper understanding of everyday experience.

Evidence presented in this chapter supports the position that human brains change in response to the environment in which they grow especially the social interaction and technology experienced by the brain. A growing body of evidence supports an idea scholars refer to as the social construction of knowledge, which posits what a human knows (skills, content, attitudes) and how a human knows it depends on the conditions under which the human learned it. Interactions among humans in large part determine what and how each individual constructs knowledge and how that knowledge can be used. Accordingly, knowledge can accurately be considered a phenomenon that emerges from human interactions, and in the absence of those interactions, learning is limited and knowledge is degraded.

The Technology Landscape

The dominant information technology has been print for so long that it has been natural for generations of students and teachers to conclude skills using text should be the focus of what we do in schools. Schooling today occurs within and prepares students for a society of dynamic information in which powerful cognitive tools are ubiquitous. Many of today's educators were adolescents when print still dominated culture, but print is being replaced by digital electronic information and this is necessitating new skills.

Information Overload

As processing speed and memory (both random access memory and read only memory) of desktop computer systems expanded and as broadband network connections became available, the amount of information available via network connections grew in quantity and expanded in type. Today, an individual with access to a personal computer and an Internet connection has access to databases that index thousands of periodicals and a constantly updated and expanding collection of information, including audio and video, is immediately available to users around the globe.

How Much Information is a project sponsored by The Global Information Industry Center at the University of California at San Diego; researchers in that group periodically quantify the media consumption in the United States. In 2003, researchers with the project concluded that five exabytes of information had been created in that year. Although that number is based on assumptions that are difficult to confirm, and the interpretation of that amount of data is equally difficult, the researchers estimated that five exabytes is approximately equal to all of the words ever spoken by every human. As impressive as that amount of information appears to be, it seems tiny compared to the amounts of data described in the 2009 edition of the study, which begins:

> In 2008, Americans consumed information for about 1.3 trillion hours, an average of almost 12 hours per day. Consumption totaled 3.6 zettabytes and 10,845 trillion words, corresponding to 100,500 words and 34 gigabytes for an average person on an average day. A zettabyte is 10 to the 21st power bytes, a million million gigabytes. These estimates are from an analysis of more than 20 different sources of information, from very old (newspapers and books) to very new (portable computer games, satellite radio, and Internet video). Information at work is not included (Bohn and Short 2009, 7).

Tucker (2013) cites data indicating the amount of information created in 2012 was 2.8 zettabytes, and that in 2015 that amount was expected to double. In late

2014, statistics provided by YouTube indicated there are over 100 hours of content uploaded to the video-hosting site each minute.

In this environment of almost indescribable and inconceivable amounts of information, users of digital media must deal with the stress of having access to too much information. Palfrey and Gasser (2008) suggested that information overload that marks the 21st century information landscape arises from several characteristics of the 21st century information landscape. Information is digital, thus it can be copied at near zero marginal cost, and the copies are indistinguishable from the original.

Also, many organizations have redefined strategic and logistic goals to make use of the vast information that is available. As is typical with technologies, it is unclear if the goals were defined first and the information was created to accomplish those goals or if the existence of the information motivated new goals be defined. It is clear that organizations are modifying goals in response to the information and the information exerts active influences on the goals of many organizations.

In addition, the information that is available appears to be more sophisticated than it was previously. Steven Johnson, a well-known writer about popular culture and the influences of information technology on popular cultures argues that television and other media (including video games) are becoming more complex with richer narrative, more characters, and more complex plot twists. Compared to 20th century media, Johnson (2006) observed the modern media landscape comprises "Games that force us to probe and telescope. Television shows that require us to fill in the blanks, or exercise [our] emotional intelligence. Software that makes us lean forward, not lean back" (136). He concluded that consuming media in the 21st century causes a cognitive workout in a way that 20th century media did not.

Ubiquitous Technology

Information technology has penetrated much further into young peoples' lives than it did in previous generations. For the adult who grew up with a phone that was attached to the wall of one's home and shared with all members for the family, the concept of each child having a phone for his or her own exclusive use that is always in his or her pocket seems unnatural. The digital generations live in a world where it is natural for everyone to have and control a device to connect to information and interaction all the time.

Smart phones that have many tools in one device seem unnatural to adults whose phones and cameras and notebooks were different devices. Because convergence is especially observed in handheld devices, all of these information tools are becoming ubiquitous and can reasonably be used applied to cognitive tasks in the classroom. Students have the capacity to take a picture of notes and diagrams drawn on the board, they have calculators, and they have the capacity to post notes on the web using the devices in their pockets. Schools that have adopted bring your own device (BYOD) approaches to computing allow students to connect their own computers to school networks, thus the school leverages the computing capacity in students' bags and pockets.

Christine Greenhow, a scholar from the University of Minnesota, observed that the digital generations are active learners in informal settings and that computer and information technology plays an important role in that learning. This combined with the observation that dissonance results from differences between everyday experience and schooling; the work of embedding technology into schooling represents an important challenge for 21st century educators. Greenhow stated the challenge for 21st century educators is to answer these questions:

> Can educators similarly tap students' enthusiasm and creativity to shape and carry out their education agenda? Can educators link students' in-school learning and out-of school living to make education more relevant, meaningful, and connected to kids? Can we bridge conventional schooling practices (where content, in many ways, is centrally determined) with informal learning practices where [users] spontaneously create content and share? (2008, 188)

Meeting these challenges will require educators adopt and adapt strategies that reflect the emerging understanding of the nature of human knowledge and the role of technologies in human interaction.

Whereas the information skills necessary for using print were largely taught through formal instruction, the skills for using 21st century information technologies are being learned organically and informally. Because devices and necessary skills change so rapidly, those who are quick learners about technology tend to be the most skilled users of it. Also, those who persevere when they encounter difficulties using technology and those who are resilient after difficulties tend to have higher levels of satisfaction when using it; that satisfaction is associated with a positive affect towards technology and higher skill with technology. These effects are largely attributed to and contribute to self-efficacy; those with these habits perceive themselves as competent and are thus confident.

The emerging sophistication of digital media and the accompanying sophistication of media skills associated with ubiquitous technology are captured in the observation of Seels, Fullerton, Berry, and Horn (2004) that interest in and attention to media is characterized by a bell-shaped curve (see figure 3.5). Media that are familiar, simple, redundant, and expected are associated with low interest and low attention because they are perceived as boring. Media that are too novel, complex, inconsistent, unpredictable, and surprising are associated with low interest and low attention because they are perceived as incomprehensible. Between these extremes of low interest, there is a level of novelty and complexity that creates a high level of interest and attention. The authors argue that as users of media gain experience, more complex and novel media are required to hold their attention, so the location of the curve is moved to the right. As all users (including young people) gain experience with technology and complex ideas, learners will demand new and more complex ideas and media in classrooms at a rate greater than expected by populations in previous generations.

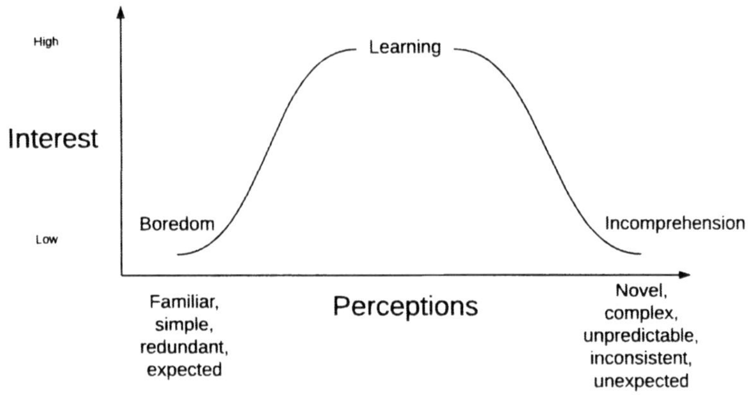

Figure 3.5. Interest in media versus users' perceptions (adapted from Seels, Fullerton, Berry, and Horn 2004)

In the 21st century, educators understand the brain as a plastic organ that is designed to learn in social situations. They also understand that the technology experienced by the brain exerts an active influence on what it learns and how it learns. Designing curriculum and learning experiences to reflect the technology and goals of print-dominated society is no longer a tenable approach.

Chapter 4
Learner Tasks

Classrooms are places in which teachers assign tasks to students (who are youngsters or otherwise uninitiated in the subject). These tasks are intended to:

- introduce the knowledge central to the subject of the course
- provide guided practice as the students learn and experience the curriculum
- provide a venue through which students demonstrate proficiency or expertise in that knowledge, including those educators use to evaluate students' learning

Throughout the industrial age and into the information age, most of the knowledge and skills necessary to be literate and numerate were relatively known and stable. For most professional educators, those skills were developed as they completed undergraduate studies to enter the teaching profession and those skills were honed and refined as they continued their careers. Classrooms became places in which they transferred parts of their knowledge to the students. They structured activities and made use of textbooks and other media and technologies to accomplish that goal and proceeded through the activities in the order captured in the list above.

Douglas Thomas and John Seeley Brown (2011) concluded these classrooms were characterized by mechanistic education, which focused on declarative and procedural knowledge. Educational psychologists Ronald Gallimore and Roland Tharp (1992), captured the nature of these classrooms,

> For over 100 years, there has been ample evidence that recitation, not teaching, is the predominant experience of American school children. Sitting silently, students read assigned texts, complete 'ditto' sheets, and take tests. On those rare occasions when they are encouraged to speak, teachers control the topics and participation (175).

In this recitation script, we observe the dominant role of the teacher in controlling the curriculum as well as the dominant role of print as the information technology.

Efforts to improve curriculum in these settings focused on identifying new information that needed to be delivered; efforts to improve instruction in these settings focused primarily on improving the efficiency of information delivery. Efficiency is measured in more texts being read, more sheets being completed, and more tests being completed with higher scores. Almost all of the educational technologies introduced in recent decades (along with new instructional models, curriculum standards and accountability systems) have been promoted as improving one or more aspect of this system. Rarely have advocates questioned the nature of the literacy and numeracy students are expected to develop as a result of the schooling experience.

Consider the overhead projector as an educational innovation. Prior to the overhead projector, a teacher wrote notes and drew pictures on slate boards with chalk. When writing or drawing, a teacher's back was to the students and the contents were bounded by the edges of the slate. With the overhead projector, teachers faced the students when writing or drawing on acetate sheets with markers of different colors. An endless supply of acetate sheets could extend the surface, and sheets could be prepared prior to the class and sheets from other sources (such as textbook publishers) were available, so sheets could include professionally prepared illustrations and edited text.

Overhead projectors and acetate have largely been replaced with computers connected to projectors, so teachers show text, images, and video coming from digital sources. Through three generations of technology, the experience of curriculum and instruction has been unchanged for the student as they sat and consumed (and continue to sit and consume) the words and images on the wall, albeit with greater amounts of information presented in a more efficient manner.

With the arrival of digital electronic computers late in the 20th century, the stability and predictability of necessary literacy and numeracy skills and knowledge evaporated. The ability to know information and to be able to recall it on demand is a skill that is losing importance as online encyclopedias become available on handheld computers. Calculators on the same handheld devices are decreasing the importance of being able to perform mathematical algorithms with pencil and paper. Those tools make increases in efficiency unnecessary, but necessitate different computing skills and extend capacity for applying algorithms.

The same technologies are increasing the importance of skills for finding and using information (such as participation, remediation, and bricolage), while remembering is losing importance. The global connections that are now common increase the importance of cultural awareness as a skill for 21st century students. The speed at which digital technologies are evolving is increasing the need for individuals to be able to adapt their information skills in an equally rapid manner. For these reasons, information skills for print are being replaced.

In his popular 2005 book *A Whole New Mind*, Daniel Pink concluded that the linear skills and thinking necessary for 20th century industrial and informational workers that were well-taught by 20th century curriculum and instruction will no longer be sufficient for success in a global economy of abundance in which many of the tasks previously done by humans are now done by computer technologies. In a report focusing in the academic skills necessary in the 21st century workforce in the United States, Jill Casner-Lotto and Mary Wright Brenner (2006) observed that the 21st century economy requires workers who have basic knowledge such as those developed following the recitation script, as well as the ability to apply that basic knowledge to unexpected problems. In the 1992 book *The Work of Nations*, Robert Reich suggested the basic work skills necessary for 21st century worker would include:

- *abstraction* which includes the ability to make meaning of complicated and unfamiliar situations
- *system thinking* which includes the ability to deconstruct the abstractions, and make logical predicts and develop rational strategies
- *experimentation* which includes the ability to drawn reasonable and evidence-based conclusions to improve the systems that drive thinking
- *collaboration* because emerging problems are so complicated that broad expertise is needed to devise and evaluate solutions

In this chapter, I construct classrooms in a manner that will make use of technologies and support students as they develop uniquely human capabilities to learn, so there is opportunity for students to gain experience developing basic skills Reich sees in their future. As we will see, the information landscape changes the rationale of education and how teaching and learning should be structured. Early in the 1940's, biology was a science being revolutionized by the discovery of deoxyribonucleic acid (DNA). English biologist D'Arcy Thompson commented, "We have come to the edge of a world of which we have no experience, and where all of our preconceptions must be recast" (cited in Gould 1998, 404). In many ways, Thompson's metaphor applies to educators who are seeking to reinvent learner tasks. The tasks we design for learners and the role of the educator in supporting students in those tasks are ones with which we have little experience and where our preconceptions must be recast.

INFORMATION TASK LANDSCAPE

The cognitive tasks that can be reasonably placed within the domain of education are broad and diverse, and increasingly generalized. Some can be easily and efficiently completed with digital information technology; others must be completed with the application of human capabilities. These differences will be the basis of educators who make reasonable decisions regarding curriculum, teaching, and learning.

Capabilities of Technology

I used to recoil when the term "leverage" was applied to computers in educational settings; I had heard too many administrators and vendors describe how some tool could be "leveraged to improve student outcomes." As we chatted while waiting for a meeting to begin, an English-teaching colleague pointed out that the word really does apply to educational technology. During the meeting, I envisioned the three classes of levers I had taught to middle school physical science students many years earlier, and it came clear to me. Digital electronic and networked computers and devices can ease many of the tasks we ask students and teacher to perform and can be used to perform complex tasks that are otherwise not possible in the classroom.

As long as we understand computers and select the problems and aspects of problems well suited to the capabilities of technology, then humans can gain advantage by using computers for cognitive tasks that would be difficult for us to complete without technology, thus we leverage technology for our purposes. In 21st century vernacular, we can download certain tasks to our devices.

Listening and Shouting

Two capabilities of networked computers that can be anthropomorphized are listening and shouting. When humans listen they attend to the sounds in the air and attempt to differentiate meaningful sounds from noise. When humans speak or shout, they can communicate with those who are listening. This system requires only the anatomy and physiology of the human body and speech develops in most humans with no special training. Speech and hearing are ephemeral and local, however. If speech is unheard, then it is lost, and it functions only over relatively short distances.

Humans can extend the distance and time over which speech and listening can be effective, but the processes are called writing and reading. Messages written on paper are much more permanent than spoken words and they can be easily transported. Print extends the audience of written speech as many identical copies can be made and distributed. Writers are communicating with readers who are situated in the future, so reading and writing can be interpreted as communication through time. With writing and print, the speed of communication is limited by the greatest speed at which the physical objects can be transported.

The earliest forms of electronic communication effectively annihilated time and distance as relevant factors in human communication. Telegraphs carried messages across continents (and then oceans) with unimaginable speed, with the only significant delay arising from the humans who encoded and decoded the messages in a series of taps. The term "global village" has been applied to capture the observation that with electronic communication (from the first telegraphs to 21st century satellite-mediated telecommunications) everyone on the planet can be within earshot of everyone else.

The listening and shouting capabilities of 21st century technology can be understood by considering the two billion smartphones carried by humans. With mine, I can send messages as text, image, audio, or video to individuals who I select or to one of several servers on the World Wide Web, thus share the message with anyone in the world (who has the capacity to find and read it). I have watched sporting events with friends across the continent as we sent messages back and forth as we watched the same broadcast. I have updated my boss on system upgrades using video chat between my office in New England and his hotel room in Finland. I have sat in a conference room in Texas and been part of a group conversation with individuals in New York and California. These illustrate the capacity of computers to shout across the globe that has become commonplace in the 21st century.

My phone is equally adept at listening for messages. There are uncountable email, text, social media, and other messages traversing the Internet at any moment. On a typical day, several hundred of those are intended for me. I have configured my devices to listen for those and to notify me when they arrive (sometimes in a subtle manner, sometimes in a blatant manner). My phone even pays attention to the sender of the message and can vary the notification depending on the identity of the sender and the nature of the message. Electronic devices listen with much more precision and with much greater patience and energy than any human. My phone listens for messages delivered when I am asleep or otherwise occupied, and it listens to many channels at once.

The capacity for global shouting and listening provided by information technology can be leveraged for many purposes in education. Educators and students can share details of classroom events, they can learn about events just as they occur, and they can extend the classroom community to mentors, experts, and peers beyond the limits of the campus.

Memory and Speed

Perhaps the most impressive capacities of computers are those related to information manipulation and analysis. Computers can remember with precision and longevity, and computers can follow algorithms at billions of steps per second.

The keyboard stokes that become digital displays that humans recognize as words and sentences are actually a series of digital signals. Those signals are stored as magnetic signals or optical signals on disks or electrically in memory. As long as the physical media are safe and the file is not otherwise compromised, the messages can be saved with precision indefinitely. If an appropriate system to read the files is available, the file can be recreated as well. It is not unusual for computer users to find disks misplaced for years and to open the files exactly as they were created years previously.

Related to the capacity to remember is the capacity to copy; once a digital file is recalled from memory it can be copied with a few clicks. The copy of a digital file is identical to the original, so the fidelity of digital information is not degraded as copies are made and as copies of copies are made. Further, copies can be made with a marginal cost that approaches zero; once a computer system is purchased and powered and it contains sufficient space to store the file, saving copies of files adds nothing to the expense of the system.

The effectively infinite memory of computers can be applied for many tasks in education. A virtual classroom can provide an archive of a student's experience. Students enrolled in Advanced Biology (for example) can access curriculum materials and assignments that were part of Introductory Biology to review as necessary. The details of homework assignments can be accessed online indefinitely (at least until students or their parents realize missing work is threatening to result in a student having a failing grade on a report card).

Computers follow rules with precision and speed; computer processor speeds are generally measured in gigahertz which are billions of operations per second. As a result, computers have amazing capacity to perform tasks that can be broken into clear steps can be expressed as algorithms. Most of the mathematics included in school curriculum is built upon problems and procedures that are algorithms. Processors performing billions of calculations per second can draw and update graphs in fractions of a second and they can perform complex statistical analyses of large data sets in similarly brief periods. Using technology to manipulate information, teachers can open interesting new ideas to students that were previously unapproachable because of the complexity of the calculations. Fractals (for example) are complex and beautiful patterns that have wide applications to many problems. Using computers, students can manipulate and explore fractals in a manner that is not possible without the devices.

In addition to performing mathematics, the capacity of computers to perform billions of steps per second can be applied to other educationally valuable tasks. For example simulations of scientific and medical phenomena and even games requiring complex strategies allow students to explore complex and even dangerous activities from safety. In addition, sophisticated games and business simulations give students experiences and allow them to explore strategies that are not otherwise possible.

The vast amounts of information available on networked systems are useless without efficient search and sort functions. The capacity of computers to perform algorithms quickly and precisely is used to identify documents that contain relevant text and to filter large databases so students and teachers can easily navigate digital information.

Mobility

One of the most obvious changes in computing devices in recent decades has been the decrease in size. In the 21st century, computers have completed the transition from desktop devices to hand held devices. Smartphones can be configured to preform many functions that once required powerful computers, and advances in battery technology allow many hours of smartphone use without recharging.

The smart phone I carry has about 100,000 times more computing power as my first desktop computer (64 GB in my phone compared to 128 kb in my first computers), and its retail price is about one tenth the price of my first computer (about $200 compared to $2000, and this assumes 1986 dollars have value equal to 2014 dollars). With wireless and cell phone networks available, I can almost always find a connection to the vast information resources on the Internet as well. My first computer arrived on my desktop about 10 years before the Internet became available to consumers, so the only data in my computer were those that I created or those that I loaded using floppy disks.

An interesting situation that I encountered when revising this chapter illustrates the power of the mobile devices that students are carrying into schools. A school was experiencing serious network disruptions, so there was effectively no Internet connection for students and teachers. Troubleshooting and repairing the problem took more than one week, and students were preparing for the end of the term. So their teacher could download and comment on essays they had uploaded to the virtual classroom and so she could upload study guides and links to relevant online resources necessary to prepare for the upcoming exams, students created several personal hot spots using their cell phones. The teacher and students who had limited Internet access away from school were able to access the files they needed using the personal hotspots. The students used to capacity of their mobile phones to bypass the malfunctioning school network.

Capabilities of Humans

Clearly, computers can be used for many educationally relevant purposes; a modestly priced smartphone can support learners as they access and manage, analyze and manipulate, create and share information. In many of these functions, the computer performs cognitive tasks once performed by humans, so we can conclude the computer replaces the human. Computers cannot, however, replace all aspects of human cognition.

An important aspect of navigating the information task landscape is recognizing the difference between tasks well suited for technology and those well suited for humans. Well-designed curriculum will provide learners with experience applying technology as appropriate, developing and refining the human-only capacities, and understanding of the differences.

Philosopher and scientist Michael Polanyi (1966/ 2009) used the term tacit knowledge to describe understanding that is implicit and difficult to state with precision. For this reason, tacit knowledge cannot be stated as an algorithm, so it cannot be downloaded to digital devices. According to Polanyi, tacit knowledge is necessary to frame a problem, to develop a strategy for solving it, and to predict and evaluate the outcomes of solution. Humans alone have the capacity for tacit knowledge, and they develop it through experiences that Polanyi calls indwelling. Through indwelling, explicit knowledge is deconstructed and then reconstructed by the student while modeling the actions of the teacher (and other experts) and by reflecting on the knowledge and its application. It is through well-developed tacit knowledge that humans understand which problems to solve and how to apply the capacity of technology to solve problems and how to interpret ambiguous results.

Wisdom

For more than 25 years, knowledge management has been organized around a hierarchy. According to the data-information-knowledge-wisdom (DIKW) model traced to Russell Ackoff in 1989, *data* comprises symbols representing the facts. Data become useful *information* as it answers questions. Information becomes *knowledge* as it is organized into generalizations and can be used to explain answers. *Wisdom* arises when humans evaluate answers, especially with ethics and aesthetics. In the hierarchy from data to wisdom, there is also a continuum from neutral to non-neutral.

The temperature outside is an example of neutral data that can be created, stored, and transported via networked computer. With no context, that data has little meaning and it can be reported precision that can be objectively evaluated. Temperature becomes information as it is placed into context. For example, we may judge 65° F to be a warm day or a cool day depending on the season and one's location. That information can become knowledge of climatic tendencies. A human can evaluate that temperature to make judgments about climate change as she creates and demonstrates wisdom. A computer may be capable of storing data indefinitely and manipulating it to provide answers to our questions, but only a human can decide if the information should be stored or what actions are ethical given the answer.

In his 2010 book, *Wisdom*, the award-winning writer about science and society Stephen Hall (2010), noted his interest in the topic arises from the question, "How do we make complex, complicated decisions and life choices, and what makes some of these choices so clearly wise that we all intuitively recognize them as a moment, however brief, of human wisdom?" (6). Hall recounted the story of a scholar who has become a leader in the field of wisdom studies; his paraphrased his colleague who observed:

> Wisdom represented a state of mind beyond standard metrics of intelligence, and this revelation forced him to see inherent failures in the educational system, and the philosophy of educational testing, and the degree to which too narrow measures like IQ tests fail miserably to predict lifetime satisfaction (245).

From Hall's observations, it appears that recognizing and defining wisdom cannot be made explicit, but emerges only from experiences in which tacit knowledge is developed and applied.

According to Hall, there are eight pillars positively associated with wisdom; individuals with greater capacity for these are more likely to be regarded as displaying wisdom:

- emotional regulation
- knowing what's important
- moral reasoning
- compassion
- humility
- altruism
- patience
- dealing with uncertainty

In these, we see a list that is in stark contrast to the capabilities of computers; except perhaps in patience, although human patience and the patience of computers to follow directions are different meanings of the term. Only humans can understand what should be done. Only humans can assess the effects of information on other humans. (This is the one human capacity that appears to be the most threatened by computers as this is generally done through heuristics in humans, and heuristics can be finely tuned and made sensitive to many channels to which computers can listen with infinite patience and super-human precision.) We see as well in wisdom the uniquely human capacity for managing paradox in both social and information situations.

Extending this reasoning, Hall concludes that wisdom is knowledge and skill that can be applied in unpredictable and diverse circumstances. The problems that face humanity in the 21st century, Hall concludes, will require people who have experience and wisdom that is aligned with the eight pillars rather than people who can succeed in classroom that were typically found in the 20th century. As educators plan educative experiences so that students have experience leveraging technology for cognitive tasks otherwise unavailable, they will also be designing curriculum to give students experiences developing wisdom.

Adopting a Naturalistic Stance

Computers can solve only those problems that can be reduced to an algorithm (artificial intelligence researchers are working to change this, but currently, computers require algorithms). Reduction to an algorithm requires rules to be clearly and completely defined, as cause and effect must be clearly specified as parameters. Algorithms are also context independent, so they can be used for every situation in which they are appropriate. Consider a program to find roots of quadratic equations. It can be used to solve every equation that is a quadratic, but if one attempts to solve a cubic equation with the algorithm for a quadratic, then the algorithm will fail. The failure can cause one of several outputs. Depending on the cleverness of the programmer, the program may simply stop working, it may report an general error, or it may identify the problem that caused it to fail.

Philosophers of knowledge describe problems that can be solved by an algorithm as positivist. Philosophers of knowledge contrast positivist knowledge with naturalistic knowledge. As the name suggests, naturalistic knowledge more closely resembles knowledge as it is encountered in the "real world" or in natural settings.

Positivist approaches assume there is a single reality and a single and universal solution for solving problems within that reality. Naturalist approaches recognize multiple realities that are constructed by individuals. Different and contradictory realities can be both logically sound and supported by evidence. Because realities are constructed, what is known cannot be separated from the person who knows. Different individuals understand and interpret the same information in a slightly different manner, and through discourse we come to more accurate understanding of what others know and how others know it.

Whereas positivist approaches reduce problems with clear and well-defined causes and effects that have clear boundaries, naturalist approaches recognize causes and effects are not clearly bounded and the connections between them are incompletely and inaccurately known. In addition, naturalistic approaches recognize there are probably unknown causes and effects. Because of this, positivists seek to define generalities that can be used in many situations, but naturalists treat each situation as unique. Naturalists also recognize knowledge as value-laden; expectations, context, and emotions all influence what is known. Clearly, different problems require different approaches; it is unreasonable to approach solving a quadratic equation from a naturalistic perspective. It is equally unreasonable to seek to solve a wicked problem through positivist approaches.

EDUCATIVE EXPERIENCES

At the start of chapter two "Education is a Wicked Problem" the purpose of education was recognized as a subject of debate as societies negotiate what happens in schools. The dominant assumption of participants in these debates is that education should "prepare students for the future." In contrast to deeper learning, which posits students must find connections between the curriculum and their lives, instructionism proceeds regardless of students' perceptions. During instruction, students may not see the connections between what they are learning and life, but teachers understand the connections, and learners are expected believe teachers and others who identify the lessons as important.

For John Dewey, the American philosopher who studied problems in education with particular attention, this was not so. Dewey posited educative experiences must build upon what learners already know (or think they know), their lessons must become the foundation for further learning, and the experiences have social value. Above all, Dewey argued, education must be perceived as immediately relevant to the learner.

Dewey's vision of educative experiences has been supported by discoveries in the learning sciences since he worked in the early 20th century. We have encountered deeper learning as explicated in *How People Learn* (National Research Council 2000). Further, Hanna Dumont (2010) and other scholars associated with the Centre for Educational Research and Innovation, which is part of the Organisation for Economic Co-operation, and Development, identified emotion and motivation as gatekeepers of education. Based on their meta-analyses of the learning sciences, they concluded only that information and those experiences that are perceived to be relevant to students will receive sufficient attention to be the focus of educative experiences. Educative experiences are further designed to encourage knowledge building as opposed to information recall and they are designed to facilitate the development of multidimensional capacity for further learning by students.

Knowledge over Information

As has been established, the instructionism of the 20th century was designed to transfer information into student's brains. Educative experiences necessitate a more sophisticated purpose for teaching. Data and information, which are the focus of instruction, are still needed, but alone those do not represent adequate education. Classrooms designed to support understanding that proceeds from information to knowledge and wisdom are more effectively facilitate students' learning in a meaningful way.

Whereas instructionism leads to learners knowing *about* the subject, Marlene Scardamalia and Carl Bereiter (2006), educational scholars from Canada, suggest knowledge-building leads to knowledge *of* the subject. Douglas Thomas and John Seeley Brown (2011) suggested that knowledge building occurs as learning *within* the subject. Regardless of the preposition used to describe it, knowledge building has characteristics that differentiate it from instruction. Learners who engage in knowledge building apply elements of human wisdom when they articulate generalizations and create new understanding; this enables the transition from information to knowledge and wisdom.

According to Scardamalia and Bereiter, "All understandings are inventions; inventions are emergent" (2006, 103). Emergent properties are those that appear only when systems achieve a certain level of complexity, and are not predictable even if the parts of the system are well known and understood. Because of this, educators who create curriculum focused by knowledge building seek to design classrooms that allow complexity to emerge through use consumption and creation of information and interaction.

A common example used to illustrate emergent properties is the well-functioning office to illustrate the emergent nature of knowledge. Writing for business audiences, John Seeley Brown and Paul Duguid observe, "Most systems, amalgams of software and hardware from different vendors, rely on social amalgams to keep everything running" (2000, 77). Expertise in each part of the technology system in an office is distributed across the individuals, and it is only through the interaction of these individuals at the correct time, that the system functions in an optimal manner. Complete knowledge of how to operate the office is an emergent property that is observed only when the parts are intact and interacting.

Classrooms designed for knowledge building and classrooms designed for instruction approach authoritative resources in different ways. Whereas the instructionist will defer to an authority assumed to possess the correct answer, knowledge builders seek to verify through evidence. When knowledge building, it is unusual for students or teachers to justify conclusions "because the textbook says so;" information from any source is understood as "work in progress" and ideas are always subject to improvement. In these characteristics, knowledge building is supported by technology. Though networks, educators and students can access multiple primary resources, including those from disparate perspectives. In addition, understanding that is created in a digital file can be revised with no effort to recreate the original, so documents can be updated and improved as understanding improves in a manner not possible with print resources.

In addition, knowledge building necessitates more sophisticated goal setting than is common in instructionist classrooms. Bereiter (2002) differentiates three types of goals used frame and motivate learning tasks in classrooms. Task completion goals are behavioristic; students may be completing the tasks as specified, but there is little attention paid to the purpose of the tasks or the use of the lesson beyond the immediate need to complete the task. Learning goals are characterized by student's engagement to a greater degree than simply completing the task. The purpose is clear and students seek to understand the lesson in his or her mind. The purpose of knowledge building is developing skills applicable beyond the immediate need; students understand the learning as it can be applied to the present situation and to other similar and novel situations.

A single activity can comprise all three types of goals: A student may propose a science fair experiment by answering a series of questions posed by the teacher to ensure the project is safe and reasonable; this would be organized around a task completion goal. Completing the project would require the student reach several learning goals that are related to understanding and answering the question that focuses the project. Presenting the project to peers, parents, and community would be focused around knowledge building goals. In presenting, the student sees communication as a valuable skill and also perceives the connection between his or her project and others' projects.

Capacity for Learning

Through instructionism, educators seek to help students achieve. Achievement is typically presented as metaphorical bar that one hopes to clear by scoring at or above a certain level on a test. The score is presented and interpreted as a representation of the student's learning and understanding. Once a goal has been achieved, attention is turned to other goals; it is an unstated and probably invalid assumption that a similar level of performance could be achieved later and the learning demonstrated by the level achieved on the test can be applied to other situations.

As described by John Dewey, educative experiences are those that prepare learners for further learning. As learners explore relevant problems in knowledge building situations, they develop multidimensional capacity for future learning; the dimensions include both explicit skills and knowledge and tacit understanding. This capacity is developed, practiced, and refined (but not achieved) as students experience the curriculum. The dimensions of capacity are boundless in that most individuals can perform at some level, but there is no limit to the degree to which one can practice and refine the skill. Further, the capacity developed in one area can be applied to new problems in the field or in other fields.

Ben Williamson (2013), another of the scholars associated with John D. and Catherine T. MacArthur Foundation Digital Media and Learning projects suggested that the purpose of education in the 21st century is to educate flexible specialists. These individuals will have an area of expertise, but will be able to apply the expertise in many situations and be adaptable. Flexible specialists will develop the skills they already have in response to changing needs and they have the capacity to identify and develop new expertise as it becomes necessary.

Academic Skills

My parents attended school less than 25 years before my brother and I did. Based on my conversations with them and my grandfather who taught during those years, and my cursory analysis of textbooks available when they attended school, the goals of education in those years was to help students develop what can best be described as academic skills. They and I went to school to learn to read and write, and perform calculations. We read textbooks, practiced and remembered, followed directions given by teachers, and we choose and provided answers, some short answers some longer answers. The more correct these answers, the higher the grades on our report cards. My brother and I provided enough correct answers to enter universities and earn degrees.

Many of the academic skills my parents and my brother and I developed in school are easily expressed as algorithms, and thus can be programmed into computers. The term linear is often applied to these skills; Daniel Pink (2005) and Douglas Rushkoff (1999) are among the authors who have popularized the use of this label. Linear skills are those that are very predictable; we can predict the skill necessary in the situation and we can predict the outcome of the situation.

Linear academic skills do play an important role in formal education. One cannot become educated in the sense associated with formal schooling without becoming a skilled reader, writer, and computer who is able to analyze in a logical manner. Whereas my education and my parents' education was largely considered complete once we demonstrated competence in those skills, that is now insufficient to be educated in the 21st century.

To become educated, one must develop capacity for learning. Capacity for learning comprises three elements. The academic skills that focused my education (and my parents' educations) remain an essential part of education. These skills must be developed through experience in natural setting, and learners must be competent applying the skills to natural situations with independence. Through metacognition, learners understand their skills and the application of the skills. Academic skills allow students to participate in and demonstrate learning in natural settings; natural situations give motivation for, purpose to, and context for the academic skills. Metacognition makes the learner aware of skills and applications. Without one, the others are more difficult. One need not be complete before experiencing the others, and the boundaries between them blur in educative experiences.

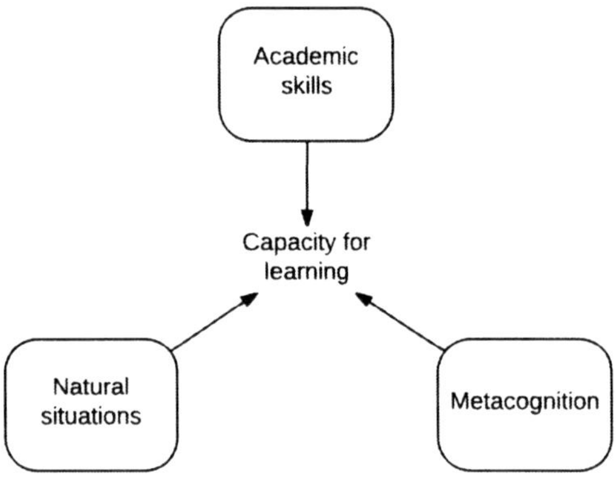

Figure 4.1. Capacity for learning comprises three elements

Naturalistic Learning

In addition to the other characteristics of instructionism, which have been well detailed through this book, context-specific learning marks that pedagogy. Many are familiar with the observation that (for example) outside of the mathematics classroom, students are less able to solve mathematics problems than they are in the mathematics classroom. Also, when asked to perform mathematics on a test, a student may score well, but when given a real world situation (even in a word problem closely related to the topic under study), the student cannot recognize the mathematics as relevant and will be less able to solve problems. This observation can be attributed to the reduction of mathematics to an activity useful for answering questions on mathematics tests; the academic skill is available to the learner only in the context of the classroom but not in natural situations.

Reduction of the complex natural problems to simplified problems for the classroom is a common approach to designing curriculum. Rushkoff observed, "This reductionist thought literally reduces the complex to manageable, if artificial, components. For many applications… this works splendidly" (1999, 19), and he concludes, "We've applied reductionist thinking to so many real-world problems that we've dangerously reduced many of our real world's problems" (19). Naturalistic learning gives students experience applying academic skills to build knowledge about complex problems rather than curriculum that has been simplified. Compared to a curriculum grounded in reduced and simplified problems, natural settings more closely resemble the expectation of educative experiences and the goal of preparing flexible specialists (Williams 2013) that Reich (1992), Pink (2005), and Casner-Lotto and Brenner (2006) claim will be necessary for workers and citizens in the future.

Naturalistic teaching and learning does exist within a continuum. A test in which students evaluate mathematics equations is an excellent example of a highly reduced task that typifies one extreme of the continuum. On the other extreme of the continuum would be naturalistic tasks. A mathematics student who is working in an office and writing and solving equations to predict productivity might illustrate the experiences on that end of the continuum. Between these extremes are word problems and cases studies, and similar materials that introduce the mathematics with accompanying situations explained or described (see figure 4.2). In some cases, completely naturalistic teaching and learning is not appropriate, but effective curriculum does provide opportunity for (and expectation that) students study complex problems.

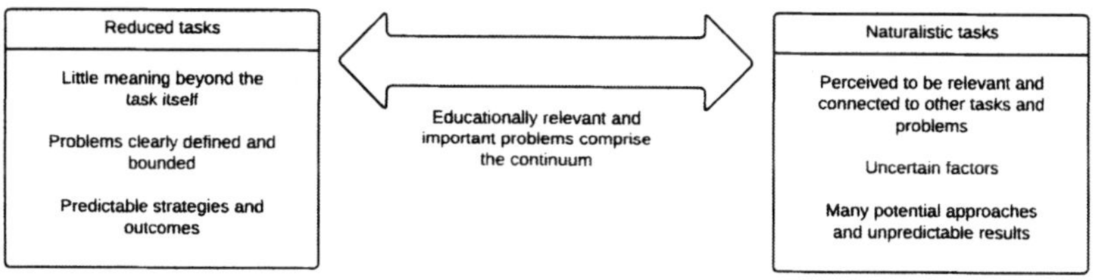

Figure 4.2 Continuum of reduced to naturalistic tasks

A heuristic that can be applied differentiate reduced tasks from naturalistic tasks is the professional recognition heuristic. If a professional would recognize the curriculum as part of his or her professional domain, then the task is naturalistic. This can also be an approximate measure of the degree to which the task is naturalistic: the easier it is recognized as part of the profession then the more naturalistic it is. Of course, we have encountered this continuum previously. When I was learning to walk after my stroke, the exercises in which I stood on pillows was a reduced task while walking on gravel was a natural task.

Metacognition

Metacognition is the process by which individuals think about their own learning. It is often described as an internal dialogue as it is a process in which the learner makes meaning of what has happened in his or her brain. While teachers can provide prompts and other tasks to facilitate reflection, metacognition is largely an individual process. Metacognition comprises several aspects of learning through which the learner becomes aware and over which the learner exerts control.

Through metacognition, learners declare what they know and become conscious of their skills and knowledge. This aspect of metacognition includes both explicit declarations of what one knows and can do and an implicit sense of self-efficacy. A learner demonstrates the outcome of this metacognition when she states (for example), "I know how to find roots of quadratic equations," and she also demonstrates it when she recognizes the path of a projectile as can be described with a quadratic equation. The declarative aspect of metacognition includes both recognition of a skill and recognition of the purpose of the skill and its connection to other problems.

A second aspect of metacognition is a learner making judgments about the sufficiency of his or her current knowledge. Those who judge current understanding is sufficient will recognize what they know and what they do not know. The incompleteness of the knowledge may turn to curiosity, which motivates further learning, or it may lead to contented and informed ignorance.

Metacognition also focuses each individual as a developing learner. Identifying interests is a small part of this; much more important is learners recognizing which methods and strategies are necessary or efficient or efficacious in given situations. Through this metacognition, learners understand how they learn, and this is developed and refined as learners gain experience learning new strategies and selecting alternatives when one proves ineffective. Metacognition leads learners to understand what they know and how they learn.

We have already seen that the technology one encounters through adolescence becomes natural and it provides the foundation for internal cognitive structures of one's skills, knowledge, and expectations (Wexler 2008). When a human finds himself or herself in a setting that is contrary to what is natural, he or she feels dissonance. Through metacognitive processes, the human reconciles the dissonance and reinterprets the situation or updates his or her internal cognitive structure. After adolescence, updating one's internal cognitive structures does become more difficult, but it is effectively impossible without the capacity to be metacognitive.

LEARNING: DEEPER, ACTIVE, AND AUTHENTIC

So far in this chapter, attention has been paid to the characteristics of information technology and human cognition, and how the capacity of both contributes to problem framing and solving by humans. Also, attention has been paid to the characteristics and products of educative experiences; those principles are the foundation for creating technology-rich educative experiences. In the current section, attention becomes more focused on the practical matter of creating teaching and learning environments for local communities. In terms of Buchanan's generalized approach to designing solutions for wicked problems, this section supports educators as they identify working hypotheses to instantiate principles of human learning to classroom experiences.

Learning science continues to be an active area of research, and discoveries are interpreted for educators and are translated in exemplary practices that more accurately and completely reflect the nature of human learning. Meta-research projects in which the findings of many studies are studied are among the most useful tools for identifying the best knowledge in a field. In learning science, those research projects are pointing to deeper, active, and authentic learning as characterizing the most effective classrooms, especially when the purpose of the classroom is to provide educative experiences.

Three terms are necessary to describe these classrooms as different research groups that are working in the field and contributing to the literature for educators use different terms, but each describes similar approaches and organizations. We have already encountered *deeper learning* as used by John Bransford and his colleagues in their report *How People Learn*. Collen Carmean and Jeremy Haefner (2002), scholars from the western United States, refer to five *deeper learning principles* that are associated with classrooms in which students more clearly understand the curriculum and are more able to transfer the skills and knowledge they learn to different situations. Jan Herrington, an educational scholar from Australia, and her colleagues have defined 10 characteristics of *authentic learning* (Herrington, Oliver, and Reeves 2007). *Active learning* (see for example National Research Council, Committee on Undergraduate Science Education 1997) has been a part of the literature and rhetoric surrounding pedagogy for several decades.

Regardless of the term used to label these strategies for teaching and learning. There are several elements of curriculum, teaching, and learning common to settings appropriately labeled deeper, authentic, and active.

Based in Real-World Problems

When challenged by students, "Why do we need to know this?" many educators respond by describing situations in which the ideas and skills being taught may be useful but only after the students have developed them. Herrington, Oliver, and Reeves (2007) noted, however, "it is not sufficient to simply provide suitable examples from real-world situations to illustrate the concept or issue being taught" (27). Students must identify the problem as important and relevant.

By leaving the complicating details intact as problems are transferred from the professional world into the classroom, educators preserve the conditions that make necessary the skills and knowledge they seek to teach. In addition, students are more likely to find connections between their experiences and the problems that reflect real world complexities. As they gain experience working through real world problems, they are likely to encounter those that cannot be solved, but the problems still retain value in the curriculum.

One of the tenants of instructionism is that problems must be simplified so that students are exposed to the essential facts in a problem. A corollary is that complex problems should be broken into small and simple steps and that only when students have mastered one should instruction proceed to he next. Ostensibly, this is done to make the curriculum more approachable and easier to understand. Scholars who have found this simplification is more like sterilization, and when over-simplified, curriculum lacks the factors that increase interest and motivation that leads students to conclude the problem is relevant. When there are fewer complicating factors, there is less opportunity for students to connect the problem to prior learning or to find an emotional connection to the problem.

Because real-world problems require days or weeks (or even longer) of effort to sufficiently understand and (tentatively) solve, authentic learning tasks should allow students to work on the problems for similar lengths of time. Artificially simplifying problems so that they fit into available time frames may be necessary, but does jeopardize the real-world nature of the problems. As with all wicked problems, however, factors beyond the control of the planner (or researcher or student) can impose limits. Curriculum based in real-world problems provides leaners experience managing solutions within this reality.

A common criticism of real-world curriculum is that students do not have sufficient skill and knowledge to frame organize, or undertake original study. While this is true, it does not provide a rationale for excluding real-world curriculum. Even when students undertake the work of verifying previously known phenomena, the continuity of their work to real and complex problems is a valid interpretation of authentic learning.

Value Competence over Compliance

In courses organized around the instructionist recitation script, the ability of students to comply with the presented knowledge and provide expected answers is the valued outcome. In deeper, active, and authentic learning environments, students who show the greatest ability to apply multidimensional capacities to propose reasonable and fact-based solutions are the most competent learners. Mehlenbacher observed of the learning environments that promote competence over compliance,

> Focus on tasks that support these processes—that is, of finding the problem, representing it, planning a problem solution, executing the plan, checking the solutions, and reflecting to consolidate learning [necessitates] tasks that move towards complex individual and social activities… (2010, 243–5).

Competence tends to be demonstrated in generalized skills that can be applied to many problems within a field of study and that can be applied to problems in other fields. Further, these skills tend to be demonstrated in products and performances rather than in testing situations. An example of these skills is the scholarly primitives (discovering, annotating, comparing, referring, sampling, illustrating, and representing) defined by John Unsworth (2000). Kozma et al. (2000) identified generating representations, accessing and navigating information, explain and analyze phenomena, and make judgments and communicate understandings as skills commonly developed by science students, and that developing competencies in these skills are more valuable than students reporting correct answers on tests.

Bloom's taxonomy and the language of higher order and lower order are used as a rationale to insist students demonstrate compliance with basic knowledge prior to students engaging in competency building activities. This is based on the assumption that the only appropriate point of entry into Bloom's taxonomy is from the bottom (through the lower order skills) and teaching should progress up through the lower order skills to the higher order skills (see figure 4.3 on the left).

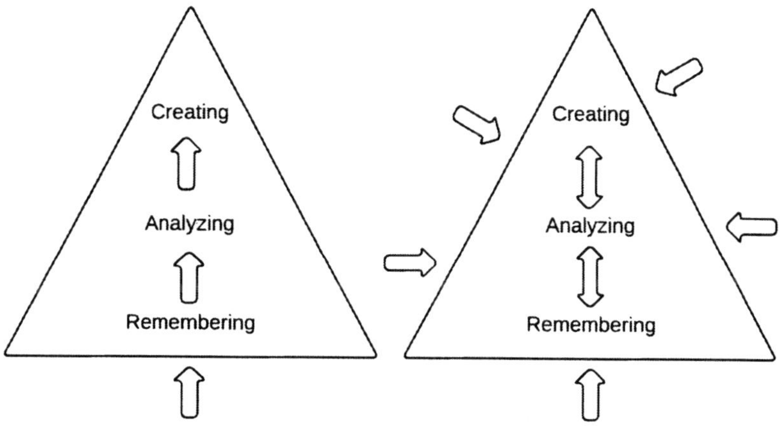

Figure 4.3. The traditional approach to Bloom's taxonomy (on the left) and a more sophisticated approach (on the right)

Advocates for deeper, active, and authentic learning tend to identify multiple points of entry into the continuum of activities inside Bloom's taxonomy; in addition, they perceive multiple and varied paths among the activities inside the taxonomy (see figure 4.3 on the right). Discoveries of the cognitive and learning support the approach of multiple points of entry into the collection of skills. Researchers cite improved motivation and cognitive engagement that results from this approach as factors that motivate learners to develop skill throughout the taxonomy.

Stance Towards Information and Interaction

Advocates for deeper, active, and authentic learning advocate for teachers to adopt several practices specific to how teachers approach information and interaction in the classroom. The exact vocabulary used varies according to the preferences of the authors or research groups, but it is clear these models articulate differences between the learning tasks in these classrooms and the instructionism that is being replaced. Specific differences include how learners approach expertise, the level and nature of social interaction, and the nature of curriculum.

Expertise

Expertise arises from both knowing about the field and experience solving real-world problems in the field. In traditional classrooms, the teacher is the individual who has the greatest expertise in the field, and hence is the community's expert. Increasingly, educators are sharing the role of the foremost expert in the classroom community.

It has been established that multitasking students who use information technology for hours each day are likely to have greater expertise using computers and finding computer-mediated information than their teachers. The teachers retain their greater expertise in using the information in a competent manner, but students have valuable information-finding skills. Just as Brown and Duguid concluded an office is a complex of the technological and social systems with individuals contributing knowledge so the whole is greater than the sum of the parts, classrooms in which teachers leverage the expertise of students are greater than the sum of the parts.

Technology also extends access to experts outside of the classroom. Through the Internet, teachers and students can access video lectures, animations, and similar presentations created by individuals with great expertise. Through the full-text database available through the school library, teachers and students have access to a far wider range of professional literature than was available in school libraries in the previous century. In addition, much reliable information is published on the Internet by government agencies, professional organizations, educators, researchers, and other credible sources. Using these resources, teachers and students can consume more resources from more diverse experts than is possible when expertise is limited to a teacher and the print resources that were provided by the school budget.

Technology also facilitates interaction with greater expertise in a manner that was not previously reasonable for most school communities. Many online forums are open to the public, and (while the terms of use may limit participation by younger students) classroom communities can participate just as the other professionals participate. Social media provides similar opportunities for classroom communities to participate in communities of experts.

Scaffolding is another method of providing access to expertise. Scaffolds come in many different forms including graphic organizers, unifying themes, and models. Each can be used to help learners see relationships between concepts and details of complex ideas. For example, a science teacher may introduce controlled experiments as scaffold in science. As they design their own experiments (a deep, active, and authentic activity), learners compare their plans to those followed by practicing scientists.

Deeper, active, and authentic does include activities such as apprenticeships and case studies for students to engage with professionals and other experts. Access to these experts provides students with different perspectives and understandings of the structure of the discipline and the methods practiced within it as applied to real-world problems. In some instances, educators who have experience in the field—for example music teachers who also have performing careers or science teachers who have worked as researchers—may be able to provide such expertise, but authentic learning requires access to both the professional educator and the expert.

Social

In the 20th century education paradigm, social interaction within most classrooms included three components (the teacher, the students, and the content), and it was dominated by the teacher and focused almost exclusively on the content. Typically, the teacher initiated each interaction, students responded, and the teacher evaluated those responses (Cazden 1987). This is the foundation of the mechanistic approach and recitation script that has been encountered elsewhere in this chapter.

For many decades, the social component of learning was recognized, but was thought to be of limited importance. Psychologist Matthew Lieberman (2013) described experiments that have demonstrated social encoding advantage in learning. The effect is unsurprising to advocates of deeper, active, and authentic learning: Two groups are given a text (Lieberman described an experiment in a which list of behaviors were provide to subjects); one directed to memorize to prepare for a test while the other told to form impressions of people who perform the behaviors. When given the same memory test later, the group that approached the list from a socially relevant perspective outperformed the group that memorized the list.

Lieberman explains the results in this way: Cognitive tasks (such as remembering a list of behaviors) and social tasks (such as interpreting people who demonstrate the same list) occur in different parts of the brain. The parts of the brain used to process social tasks create stronger memories than the parts used for cognitive tasks. It appears that when humans engage those parts of their brains associated with social learning, they develop stronger memories than when they engage those parts of their brains associated with cognitive effort.

Evidence seems to support several approaches to teaching and learning that engage the social system. By focusing attention on the social aspects of an event is one method. The result is that studying the social implications of scientific discovery or focusing on the social causes or effects of a conflict may actually improve students' abilities to remember the non-social details of the event. Encouraging social interaction focused on the details of the subject is another method. The social interaction must be more purposeful than simply working together in a disorganized manner. Especially when learners first encounter a subject, scaffolds to support their learning and protocols to facilitate social interaction can make the social learning more effective. In addition, the student who takes an active role in presenting and tutoring in a social exchange will be engaging the social system to a greater extent than the student is receiving the tutoring. For this reason, learners in deeper, active, and authentic classroom assume a more active role initiating, sustaining, and evaluating social interaction than they do in instructionist classrooms. Scholars have discovered that social interaction mediated via technology can be as effective as face-to-face social interaction as well.

Herrington, Oliver, and Reeves (2007) suggested one of the most effective types of social learning is collaboration. Collaboration occurs when the product is one that could not have been created independently by any of the participants. Collaboration leads to emergent properties which are difficult to predict, and the products of collaborative activities are often unpredictable even to the teachers who organized them. In this way, the cooperative learning protocols that are common in the literature for teachers may not produce collaboration. Further, social interaction designed to create a specific and defined product are usually not collaborative.

The problems that focus study in deeper, active, and authentic learning environments are likely to be wicked and so the multiple perspectives that different individuals and populations bring to those problems are introduced to classrooms. When designing authentic learning tasks, educators will both recognize and encourage students to consider and reconsider the problem from different perspectives, thus engaging the social learning system in the brain.

Refection

As we have seen, metacognition is an important aspect of the capacity developed in educative experiences. In deep, active, and authentic learning activities, educators provide both time and prompts to help organize learners thinking about their own thinking. Reflection can be considered an internal social interaction in which the learner engages himself or herself in discourse regarding the learning experience.

Effective reflection comprises original articulation and is facilitated by prompts or scaffolds to encourage learners to think about how the parts of their solution fit together and the processes that led them to the solutions they create. Further, they are led to revisit their process and solution to assess their strategies and approaches.

Products that Matter

In all classrooms, tasks are assigned with the understanding that they will be used to judge the degree to which each student has learned what he or she is supposed to learn. In traditional learning environments, these tasks are typically contrived and have little relevance beyond the classroom; consider the test prepared by and evaluated by the teacher; it is unlikely the test would be appropriate for any other group of students and it is a performance for an audience of one. In classrooms organized around deeper, active and authentic learning, the evaluative tasks tend to be those that professionals would recognize as products of their professional work and other performances for extended audiences.

Projects, (as defined in project-based learning, which is considered in the final section of this chapter), are a typical product of deeper, active, and authentic learning that are used to evaluate student learning. The product of the project is an original creation that requires learners to demonstrate all aspects of the capacity developed during educative experiences.

All testing is not precluded from inclusion in products that matter. In some professions—for example those related to information technology—performance on tests is used as a gateway through which professionals earn certifications or licenses. As an adult, I observed my father, who had been a truck driver for decades, studying to earn his license to haul hazardous materials as cargo. While the test covered important aspects of hauling such materials safely and in accordance with relevant laws, once he earned the license, his performance on the test was of minimal importance to his supervisors. In authentic learning settings, the products and performances used to judge students' learning are similar to those produced by professionals working in the field. If tests are presented as gateway activities (as they are in the professions) rather than as culminating activities (as they are in education), then the authenticity of the test as an evaluative task will be established.

Role of the Teacher

In classrooms where deeper, active, and authentic learning focuses students and teachers, the role of the teacher changes when compared to instructionist classrooms. Rather than dispensing information to students, dominating the discourse, and serving as the sole judge of correctness, teachers in active and authentic classrooms will serve in multiple roles.

Collins (2006) describes the teacher in deeper, active, and authentic classrooms as managing cognitive apprenticeships, which necessitates a teacher attend to several aspects of the classrooms. When managing the *content* in the curriculum, the teacher makes decisions regarding all aspects of what comprises the appropriate level of detail for students. The content decisions managed by the teacher also include strategies and materials and tasks that will provide the necessary content and scaffolding. Within the content are also organizational decisions such as sequencing the content in a purposeful manner that facilitates understanding and also coaching.

When managing the *methods*, the teacher decides the approaches that will be used to facilitate students' learning. Creating rubrics is an example of the methods that will focus cognitive apprenticeships. The teacher is making many decisions about what is important and how the learners will understand it. Finally, the teacher makes decisions regarding the *sociology* of the classroom, which determines the nature of the social interactions between students and teachers and among students that are expected. Decisions about how to frame problems, including the level of authenticity and the degree to which problems are sterilized, also contributes to the sociology of the classroom. All of these contribute to the multiplicity of perspectives engaged, the unpredictability of the products, and the level of collaboration in the classroom.

On athletic fields, coaches make close observations of the performance of individuals, and they provide advice and guidelines through both explanations and models in an attempt to improve performance. Even on team sports, coaches spend much time and energy helping individuals to improve the performance of the whole. Coaches design exercises through which the performance can be improved, but they also provide opportunities to scrimmage or practice in real situations so that individuals and teams are prepared for in competition. Educators in deeper, active, and authentic learning classrooms coach individuals and groups in similar manners.

As local experts who are always available to a community of learners, teachers are active models of the social system within which the subject is based; science teachers are models of scientists, and teachers of other subjects model those professionals. Rather than simply explaining and demonstrating (although those are sometimes necessary), educators who model assume the position of expert and attempt to simulate an apprenticeship situation for students.

Barbara Rogoff (1990), a psychologist from the University of California, Santa Cruz, proposed guided participation as a term to describe the teaching and learning that is common in pre-literate and other cultures. Guided participation proceeds from children's current understanding (what is known), and through structured activity, children learn the details and nuances of the activities and ideas. As a child becomes more skilled, he or she takes on additional responsibility makes decisions regarding his or her learning and also serves as a mentor to those who are younger. Rogoff concludes guided participation is a dominant model of learning in human cultures.

As both coaches and models, teachers also use their professional judgment to help organize experiences, prompts, and other activities to facilitate learning. Vygotsky's zone of personal development (see pages 47–48) illustrates the need for teachers to make decisions related to the appropriateness of the tasks students undertake. In addition to organizing curriculum for both individuals and groups, teachers help students to frame and limit problems. This role is particularly important in situation where students have selected their own problems to study.

MODELS FOR TECHNOLOGY-RICH CLASSROOMS

Some educators have been designing classrooms in manners that are aligned with the educative experiences envisioned by John Dewey and the other scholars whose work informed the previous sections of this chapter. This chapter ends with brief descriptions of some of the models that have proven effective in creating educative experiences that are aligned with the principles of deeper, active, and authentic learning. These models are presented as supports for the final step of Buchanan's generalized planning solutions for wicked problems. These models provide a scaffold for articulating working hypotheses regarding how the cognitive and learning sciences can be realized in local classrooms. These are models for understanding how deep, active, and authentic learning can be instantiated in local classrooms.

Project Based Learning

Just as "technology integration" has been applied to any task in which a learner is in front of a computer or tablet, "project based learning" is a term that is applied to many different types of learning tasks. For some educators, whenever the product of learning task is something other than a test, the students are participating in project-based learning. For other educators, project-based learning (PBL) occurs only when the tasks are clearly marked by essential characteristics.

To be PBL, students must exert significant choice in defining the topic, focus, questions, and purposes of the project. Because of this character, many curriculum leaders place PBL outside of the traditional subject areas. PBL organizes senior projects, middle school interdisciplinary studies, and other extra-subject learning activities. Within courses that are aligned with traditional subjects or departments, PBL is still possible as demonstrated by science fair projects and similar subject-specific research and inquiry.

By choosing their own projects, students are motivated by and interested in the work. With the guidance of teachers, the academic skills and other dimensions of capacity necessary for the project are identified and refined, and alignment with the curriculum is ensured. Students' choice extends to defining the product of the learning as well. Educators play an active role in helping students reflect on the learning and preparing it for final presentation to an authentic audience. This role is similar to the editorial and peer review processes that are common in professional communities.

Projects are designed with broad outcomes in mind. Rather than simply being an alternative method for students to report correct answers, content is deeply embedded in the project, and students make judgments about the content necessary to complete and communicate the project. Projects require original articulation by the learner and require the learner to organize and prioritize how it is undertaken and communicated. A corollary to this is the scope of a project; a project should represent significant age-appropriate performance and significant new explicit and tacit learning for the student.

Frequently, but not always, the project is connected to the local community. This local connection can take many forms, including study of a local phenomenon, service to a local population or need, or implementing a solution to a local problem. For some projects, the focus of the project is specific to the learner. For example, PBL initiatives are focused by career awareness or exploration activities and lead to apprenticeships in the community for older students.

Projects can be undertaken by all students and at all ages; they become increasingly sophisticated as students get older and students undertake projects with increasing independence. Many educators also recognize that PBL can be understood along a continuum. Valuable educative experiences can include elements of project-based learning. By including elements of PBL in the curriculum, educators create the conditions in which students develop the capacity for success in full project-based learning.

Two examples illustrate the nature of projects. First, a middle school student spent time with a neighbor who was a professional welder and metalworker during his school's 10-week PBL time. After only four weeks, he had completed all 40 hours of works and was continuing to work with his mentor. At the end of the project, he brought one of his sculptures into the school and it was displayed in the school lobby. The assistant superintendent stopped to look at the work and assumed the art was part of a program to bring works from the local art center into the school. In this example, the student had exclusive choice over the subject of his study and the nature of the performance. The audience was rather localized as only visitors to the school observed his work.

Second, high school students created apps for businesses for their computer science course. One student created an app for the towing company owned by a friend of his family. He uploaded it the online store, and a customer who needed the towing service downloaded it. That customer discovered the phone number was incorrect in the app, and the student became very motivated to update the app and makes sure his work was accurate and correct. In this project, the nature of the product was defined by the subject matter of the course and the assignment given by the teacher. The audience for this was authentic, and that motivated the student to apply his academic skill to a level he claimed never had previously.

Upside-Down Classrooms

"Flipped classrooms" is a model of technology-rich instruction that became popular in the second decade of the 21st century. Video sites, such as YouTube, had become very popular, and teachers (and those who played teachers on their YouTube channels) began uploading instructional videos and these became the focus of flipped classrooms. In a flipped classroom, the teacher points students to instructional videos that are watched prior to arriving in class (substituting for the teacher's introduction) and then the tasks traditionally assigned as homework are completed in the presence of the teacher. There are two advantages of this method:

- students control access to the instruction, so they can repeat it as necessary
- teachers are available to provide tutoring, coaching, and other guidance when students are coming to know the material

While flipped classrooms do have these advantages over a traditional instructionist method, it remains instructionist and the criticisms of in-person instructionist models can be applied to flipped classrooms as well. While students who are members of the digital generations consume vast quantities of video and other media, those same students frequently complain about the quality of video instruction, especially by inexperienced teachers, and are as likely to stop video they find unappealing, as they are to repeat video-based instruction.

Although "flipping" a classroom so that it replicates instruction does not appear to produce educative experiences, "flipping" activities other than lecture that were previously done in-person to homework does appear to facilitate the active and authentic learning necessary for 21st century learners. The term upside-down classrooms can be used to capture the flipped nature of the organization but differentiate these approaches from flipped instruction.

Natural Context through Video

While a video replacement of in-person instruction is not educative, using video to promote indwelling and other elements of tacit knowledge can be educative. When used in this way, video can replicate the informal and speech-based learning that is natural in real world experience. Through follow-up teaching methods, what is experienced and known through the video can be stated formally and in generalized statements that are new to the student. Rogoff (1990) noted this known-to-new approach is characteristic of the guided participation observed in many cultures as well. This is contrasted with instructionism, which proceeds from new (the introduction provided by teacher or video) to known (the practice or exploration provided by homework).

Because video is experienced with multiple senses, it more closely resembles real world experience than the text-mediated (or lecture-mediated) presentations that are common instruction in classrooms. In upside-down classrooms, video is used to provide students with more natural experience, which increases tacit knowledge and motivates explicit knowledge.

In my traditional schooling (decades ago), my Humanities teachers arranged for our class to view a film of "Romeo and Juliet" after reading the play. For many of us, that film caused the reaction, "Oh, that's what the story was about." Unfortunately for our understanding of and enjoyment of the play, watching the film was a culminating activity, so we were on to new texts once the film was ended. The upside down educator appreciates that value of using film and video to give students experiences and naturalistic context that makes the text-based curriculum more approachable, understandable, and relevant. Teachers who introduce poetry by showing video of poets reading their work and teachers who show video of science demonstrations included in college lectures are using video in this manner.

Data Collection

All science teachers (as well as teachers of many other subjects) are familiar with laboratory activities; students construct an apparatus and collect measurements based on prescribed methods (or students otherwise gather data). These activities are usually followed by analysis of the data to identify patterns, answer questions, and draw conclusions. A common complaint of those teachers is that students do an inadequate job analyzing the data and they draw weak conclusions. This complaint can be explained as the students are analyzing the data and working to draw conclusions (the more cognitively challenging parts of the work) when they are away from their peers and teacher, so opportunities for social interaction, support, and clarification are largely missing.

By assigning students to collect data outside of class, teachers can make time available for guiding analysis of the data. While this is a model well suited to science and math, subjects that deal primarily in quantitative data, it can be used in other areas as well. For example, my upside-down colleagues have had students analyze various texts (including both fiction and non-fiction) and analyze video (including speeches and dialogue in foreign languages). In all of these cases, the students arrive at class with their data, and reconciling differences and identifying commonalities becomes the analysis, which was guided by the teacher.

Successful data collection outside of class requires the teacher model the collection process prior to the assignment and it requires students have access to the data to be gathered. In many cases, data collection in upside-down classrooms requires the use of online simulations or other online media. Steps must be taken to provide for those students who have limited access to the Internet away from school.

Especially in the sciences, data collection in upside-down classrooms is typically done using a computer-based simulation. When gathering data using a simulation, students are not manipulating laboratory apparatus. Such physical work is essential to for students to develop tacit knowledge of science and many other fields. Because of that missing experience, teachers should plan simulations ad upside-down data collection for those topics that are difficult or impossible to complete in physical laboratories.

Reflection in Isolation

A classroom in which there is deeper, active, and authentic learning occurring will be a loud place, as it will be full of social interaction. While learning is a highly social activity, it is not an exclusively interactive activity. Reflection requires an individual to think back over the learning activities to find connections, to resolve questions, and to become metacognitive. A quiet and distraction-free environment is conducive to the contemplative thinking necessary for reflection. For this reason, upside down educators will facilitate and encourage reflection that is completed outside of the active classroom, which likely has too much extraneous cognitive load for reflective learners.

Technologies for web publishing such as discussion boards, blogs, wikis, and journals (all tools that are available through a virtual classroom as described in the next chapter), can be used to facilitate reflection. When social reflection (discussion) and individual reflection (becoming metacognitive) is mediated via technology, students can participate in settings where the extraneous cognitive load is minimal (assuming the technology does not contribute to that load).

In traditional classrooms, discussion is an important method that allows teachers to assess students' understanding and to give them experience with spoken language. In those classrooms, however, a small number of individuals can dominate discussions, some students may be reluctant to participate, and what is said is usually lost (just like all spoken language is lost). Further, the discussion is local and synchronous; those who are not in attendance cannot participate in or even observe the discussion. Using online discussions increases participation and saves the words, so they can be reviewed and reanalyzed later.

One of the common complaints of educators who begin using online reflections with students is, "They just post silly stuff." In many cases, students adopt the superficial responses that are typical of their online life away from school when participating in online reflection in school, so this observation is completely accurate. When facilitating an in-person discussion, a teacher can direct and redirect reflection to avoid distractions and seek deeper contributions. When facilitating online discussions, teachers compose a prompt that provides the impulse for sustained reflection. Just as leading discussions takes practice, so does composing good prompts.

In-person discussions are synchronous events, so educators can react to what is said immediately and steer the discussion away from irrelevant topics just as the deviation occurs or they can take steps to include those who have not participated. Online discussions are asynchronous; participation takes place at different times for different people, and so it is difficult to steer discussion in a timely manner. As a result, educators lead the discussions using different methods. During in-person discussions, the teacher sustains the discussion by evaluating what is said and providing prompts to improve the discussion and encourage full participation by everyone. Sustaining online discussions requires teachers develop and use anchors or prompts to sustain discussions with minimal direction once it starts. The initial prompts must engage participants and encourage thoughtful and extensive responses from the beginning. Online teachers tend to evaluate online prompts and anchors after the discussion and to make changes to improve it when used again. The creation of protocols that establish the cognitive structures and conditions under which students are expected to interact with others facilitates the transition from students engaged in silly interaction to academic interaction.

Teachers also reduce the "they just post silly stuff" effect by actively participating in the online reflection. Teachers can model appropriate and deep reflection, thus they share their expertise within the work of the learning community. By participating, they gain first-hand experience in the process, so can better guide students and can improve the prompts they assign.

Assessment

Educators can employ many methods to assess students' understanding of the information and students' current level of skill. In some cases, educators seek to assess students' understanding using objective measures, such as students' ability to provide or recognize correct answers to questions. In other cases, educators seek to understand and document learning capacity. Both assessment goals can be accomplished with digital tools.

Testing

Testing has been the dominant method of assessing and evaluating learners for many generations. While performances arising from complex and real problems and prepared for audiences beyond the local classroom is the best method for demonstrating the capacity students developed during educative experiences, testing will continue to be a (minority) part of classroom assessment and evaluation. Using technology to administer tests has several advantages for both students and teachers.

The great advantage of using computers to administer tests is the capacity of computers to follow rules with amazing speed and precision. Once a teacher creates a test, it can be administered and graded with no further input from the teacher. Answers are immediately marked as correct or incorrect (or partially correct) based on the rules set by the teacher. Further, once a question has been added to the question bank, it can be shared and reused, and remixed and edited with minimal effort. The result is a science department (for example) can collaborate to create shared questions, so teachers are relieved from the burden of creating, copying, administering, grading, and recording quizzes on laboratory safety which is important knowledge for all science classrooms.

Digital tests also provide educators with many options for tests delivery that are associated with effective testing. For example, feedback can be provided immediately, and test questions can include high-resolution images and other media. In addition, time limits can be applied with precision and multiple attempts based on performance can be allowed. Further, the order of question delivery and the order of correct answers in multiple-choice questions can all be randomized to minimize cheating.

Testing has become big business. Many high-stakes tests associated with the Common Core State Standards implementations are administered via a web interface. In this, the tests are following the lead of many professional organizations that administer licensing exams. (When I supported technology-rich teaching and learning in a community college, the chairperson of the nursing department explained her faculty's reliance on computer-based testing using our learning management system, "We want our students comfortable taking tests online because that's how they take their licensing exams.") By providing students with experience taking online tests, they are preparing those students for the gateway tests they will experience as emerging professionals.

Portfolios

Portfolios have long been used to document learning in certain academic and creative areas; artists in particular have maintained portfolios of their work. Increasingly, portfolios are used to document learning in all areas. As a tool for general education populations, portfolios have been found to contribute to student success, make learning visible, and contribute to learner-centered reform efforts (Eynon, Gambino, and Török 2014). Especially when students in deeper, active, and authentic classroom are demonstrating their learning through authentic performances, portfolios become an essential tool for documenting, understanding, assessing and evaluating learning.

A portfolio is the creation of the learner, thus compiling it becomes an extended metacognitive activity. There are three essential inputs into the portfolio: authentic experiences, academic experiences, and expectations (see the left side of figure 4.4).

Authentic experiences are the projects, apprenticeships, and other demonstrations of learning that are common in deeper, active, and authentic classrooms. A survey conducted by the Chronicle of Higher Education (Thompson 2012) indicated that experiences (such as internships and jobs) are more important than academics when making hiring decisions, and similar experiences are often important factors affect undergraduate admission

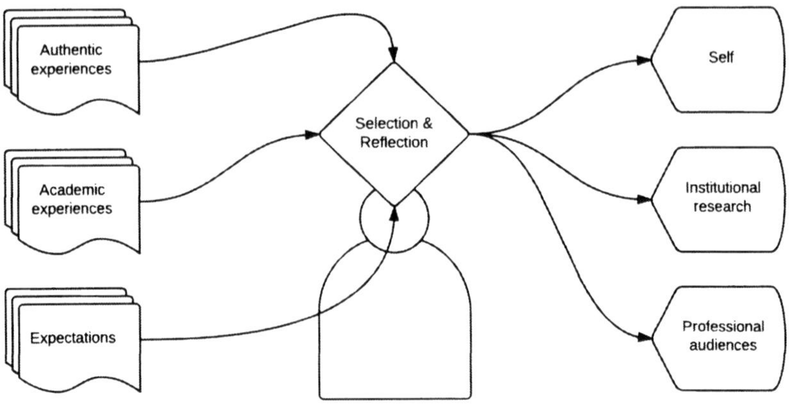

Figure. 4.4. Portfolio inputs and audiences

decisions. Through the portfolio, these naturalistic activities and the new learning that resulted are documented and enter the learning record of the student.

Academic experiences include the traditional academic skills that continue to be an essential element of the capacity that learners develop through educative experiences. When creating a portfolio, learners include descriptions of the artifacts and reflections on the artifacts, so one of the most obvious academic skills demonstrated in a portfolio is writing. Many educators discover, as well, that some of students' best writing is contained in the portfolio reflections as students find the experiences relevant and interesting, thus they care about the clarity with which they describe the artifacts and the learning represented by the artifacts.

In some cases, test scores are also included as an artifact of academic learning. This is particularly true of portfolios created for those who seek to enter professions in which tests are part of the qualifying process. Students who seek to become information technology professionals, for example, may include artifacts documenting the certifications they have earned.

The local community defines what the learner is expected to demonstrate in the portfolio. Graduation requirements, learning expectations, and vital results are all terms that have been used to define expectations. These expectations become the organizational structure for the portfolio and learners interpret their academic and authentic experiences in light of these expectations. For example, a school that includes "students will be competent writers" as a graduation requirement may encourage each student to include a page that contains examples of his or her writing. The student will describe the situation necessitating the writing, and the also reflect on his or her writing skill.

As is illustrated in figure 4.4, the learner plays the central role in selecting the artifacts used to demonstrate his or her learning and in interpreting the meaning of those artifacts. The result is a triangulation composed by the learner: The meaning of the artifacts from authentic experiences are explained in terms of the expectations. The learner then describes how the artifacts demonstrate new learning, and reflects on the meaning of the experiences. While people other than the learner (usually) define the expectations, the learner exerts complete control over the selection of artifacts and the meaning-making process.

Portfolios are a valuable source of information about a learner's experience for three audiences (see the right side of figure 4.4). A portfolio is a meta-cognitive endeavor, so the learner comes to understand his or her experience as a learner. The collection of portfolios created by students enrolled in a school can become data for institutional research. Educators can analyze the contents for evidence of students' performance and growth, thus the portfolios can inform program decisions and improvements. Finally, portfolios are used for professional purposes; the portfolio process can support all job search and college admissions efforts directly and indirectly.

Chapter 5
Understanding the Technology Infrastructure in Schools

In previous chapters, it has been established that humans are technology-using creatures, and the nature of the technology we encounter exerts strong and active influences on both our cognitive and our social development. In the current chapter, attention is turned towards the nature of the hardware and software we encounter in schools.

Technology evolves very rapidly and technology what is available on any campus is likely to have been selected and installed based on local circumstances; therefore any specifics I present here will be obsolete or inaccurate (likely both) for readers. Proceeding from the position that the technology infrastructure one encounters in a school will be a solution to a wicked problem, this chapter presents general lenses that can be used to understand problem of information technology planning, strategies for successful technology planning, and the provision of virtual classrooms which represents an emerging infrastructure need that is largely unmet in K-12 settings.

TECHNOLOGY ECOSYSTEMS

The invention of digital electronic computers during World War II signaled the beginning of a transformation in how humans interact with information; that transformation accelerated with the arrival of personal computers in the late 1970's, and accelerated even more with the opening of the Internet to general populations. This transformation has advanced at different rates for different populations and it has yet to be completed. It has been accompanied by several changes in the approach to information technology research taken by scholars. In the 1970's, information theorists adopted an information-centric view, so the reliable and efficient transmission of data was the goal of system designers. For several decades after the 1970's, researchers focused on human-computer interactions and sought to create efficient systems of precise search and efficient interfaces.

In recent decades, the third phase of information research has emerged and it has adopted an ecological character. Ecological in this case refers to the large number of interacting factors that researchers understand are influencing humans and their interactions with information and technology. As digital information emerged, they became systems used to mediate interaction between humans as well as information delivery systems. Documents (especially those editable by the public) mediated and focused that interaction in ways not previously recognized. In addition, highly customized information sources and interaction options, including highly configurable computer-to-computer interactions, are mediating the information ecosystem (Marchionini 2008).

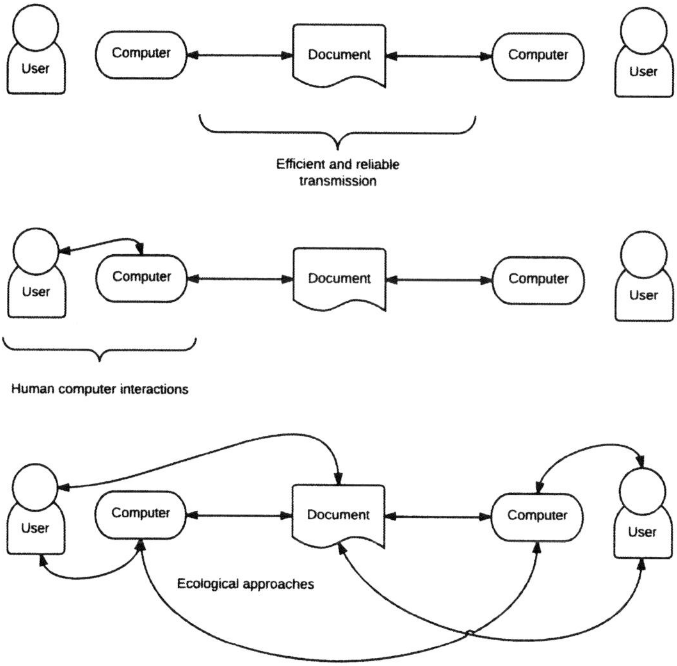

Figure 5.1: Three phases of technology research: efficient transmission, human computer interactions, and ecological

Consider the ecosystem-like interactions that characterize "live-Tweeting" during a keynote presentation at a conference. This happens when a number of people present at the speech post their reactions to the presentation on Twitter. Especially if the presenter announces a hashtag before beginning, within a few minutes, there will be a collection of posts that affirm or question the speaker, that point to relevant sources of information (including those mentioned by the speaker), and that affirm or question others who are posting. Depending on the number of people who participate and the level to which they use hashatags and connect with each other as the session is underway; the virtual discussions can take on a very dynamic nature, and occasionally direct the presentation. It is not unusual for a presenter to have an associate who pays attention to the live Twitter feed and poses questions or identifies trends that arise in the feed, so the in-person presentation responds to the virtual discussion.

In the example of live tweeting, we can identify many connected factors that necessitate we understand technology-mediated information and interaction from an ecosystem point of view. First, information objects are dynamic. While the slides used by the presenter are likely static, the presentation and the information objects appended, attached, and derived through the associated Twitter tags and feeds represent a dynamic information complex created through the insightful collaboration of the participants.

Second, the dynamic information objects are created through interaction. The individuals composing the tweets are interacting with the information and with others. The ideas that resonate with the group are strengthened through responses and retweets, and those that raise questions receive similar attention. When hashtags are used, the feed becomes more interactive as well as the devices listen for relevant hashtags and displays those the participants want to see. This aspect of the interaction is mediated thought the technology and feeds are updated with no effort from the recipient once the feed is configured. Further, in situations in which the speaker attends to the feed, the interaction includes that individual as well.

Similar observations can be made about other parts of the information landscape. A *Wikipedia* page is likely to be updated as events change, sometimes within minutes of the event, thus the information that is shared about situation or event on that encyclopedia is dynamic in a manner that information in printed encyclopedias cannot be. The comments added to YouTube pages demonstrate how the context of information is dynamic just as the known information. While the content of a video may not change, an insightful comment by a viewer can change another viewer's interpretation of the same information presented in the video. By subscribing to dynamic information objects, any user can be notified when new content is available; this interaction in between computers with no human intervention beyond the initial configuration.

The ecosystem approach to understanding and designing computer systems reflects the increasing complexity that arises from the amount of information, the options for accessing information, and the freedom that characterizes public information where the Internet is available. John Milton, in his pamphlet and speech "Aeropagetica" published in 1644, argued against the licensing laws that required permission from the government to publish. Often used as metaphor for free speech, modern digital devices represent extreme aeropagetica. Whereas access to mass media was controlled by the expense of printing or broadcasting a few decades ago, any of the two billion individuals who carry a smart phone can capture an image or video and publish it to a world-wide audience quickly (measured in seconds) and inexpensively (margin cost is near zero).

The technology ecosystem will influence how teachers create and students experience information and interaction in classrooms. Gabrielle Ivinson (2000) identified eight aspects of classroom places that can be affected by digital media and technology-mediated interactions (see table 7.1). Ivinson's list reflects the realities of the technology ecosystem in society. Information expands to be effectively infinite (much more information is available that can be consumed), and interaction extends to non-localized places and allows for more active participation by all members of the community.

Table 7.1. Aspects of classroom influenced by communication technology

Venue	Classroom Place	Virtual Space
Place	Restricted to physical location	Expandable indefinitely
Equipment	Limited to physical resources	Available indefinitely
Curriculum	Occurs at one time and place	Available from any connected computer at anytime
Time	Scheduled and synchronous	Irregular and asynchronous
Grouping	Managed by educator	Managed by educator
Space and movement	Physical	Accessed via computer
Wall displays	Limited	Expandable
Discourse	Teacher-mediated and generally limited to those present	Expanded options within and beyond the community

adapted from Ivinson (2000)

Digital Information in Schools

Schools have always been places where information is consumed and created. For most of the history of schools, that information was created as physical artifacts (works written on paper, images drawn on paper, songs recorded on tapes, and similar creations). Once physical or analog media is created, it must be copied on to bits of matter (paper and ink, vinyl, or plastic) and those physical things (books, tapes, compact discs) are protected from the elements and physically moved to disseminate the information. Obtaining a new science textbook (for example) requires one be written, copies be printed, and the copies be shipped to the students to carry and read. While making notes in the margins may be helpful to one reader, it is usually considered vandalism and may result in a student receiving a bill for destroying a resource that must be replaced. The arrival of computers in school was accompanied by the arrival of digital information; since then, teachers and students have created, copied, stored, and transported information in fundamentally different ways than did previous generations.

Digital electronic information is created as physical actions (such as keystrokes or speech with gestures) are converted into electronic signals. These are stored temporarily in the random access memory (RAM) of a computer, and then stored more permanently as signals on read only memory (ROM) disks or in flash memory. When stored in RAM, information is very easy to manipulate and transport. A few keystrokes or mouse clicks can change the contents of a file or make a perfect copy of the information, and a few more can send those changes anywhere on the globe through the telecommunications network. In schools and beyond, vast amounts of information stored in public spaces, so infinite information is now available to teachers and students.

A few generations ago, the information available to a classroom full of students might include one copy of a textbook for each student, library of (perhaps) 50 volumes of references books the teacher had collected for his or her classroom, a school library with a several thousand books, subscriptions to a few score periodicals, and comparable collections of videos and references books. Today, students and teachers have access to all of the resource previously available, as well as online references, full text databases that index thousands of periodicals, infinite online video, and the web resources maintained by credible government, media, and professional organizations, as well as academics and other credible individuals. All of these sources place vastly greater amounts of information into the curriculum than was possible with physical information.

In addition to infinite amounts of information, networked computers provide teachers and students with tools to easily and quickly search information. Given the vast information, effective methods of searching are necessary. Very rapid comparison of the contents of files to search terms permits efficient indexing, so relevant text can be identified in digital documents. Users can add tags to information sources so they participate in identifying relevant aspects of documents and make those searchable. In addition, researchers are developing search tools that can accurately interpret ambiguous search terms composed by users. All of these tools are necessary for navigate and identify useful information in the 21st century.

Technology tools can be leveraged to manipulate information as well. Complex ideas that are difficult to understand can be illustrated with digital tools. Regression lines are an excellent example. With analog information technology, students must draw "best fit" lines through data sets once they are plotted on graphs (typically paper graphs drawn by students after many minutes of effort). If the graph is drawn with a spreadsheet, however, a regression line can be added to the graph and it changes as data points are plotted. This technology-mediated manipulation of the data helps many students understand the "best fit" nature of regression line more quickly than reflecting on the slow plotting and permanent drawing of the line on a physical graph.

Finally, technology tools can be leveraged to create types of information in a manner not previously possible in most classrooms; digital information can be remixed with other information. Images and other media can be copied from sources when creating new information objects. Equipped with a digital camera (or a smart phone), users can also capture images and media that can be incorporated into information objects they create (or that can be shared for others to remix). With these tools, students can more clearly communicate, and they can use more natural channels when understanding what others mean. Multimedia capabilities also extend and enhance the communication options available to students; more students have access to more cameras and recorders that are cheaper to use than ever before.

Technology-Mediated Interaction in Schools

The events I recorded in the papers I wrote as an undergraduate student and in my journal kept during my first few years working as a teacher and the few surviving lesson plans and resource folders from my pre-Internet years (recall that I entered the teaching profession using an Apple IIc computer in 1988), all suggest my experiences reflected the contemporary trends in computing. I used my first computer to create information (especially word processor files) and to manipulate information, including spreadsheets and graphs. As computers arrived with compact disk drives and as information available on the disk became more sophisticated, my students and I consumed (and created) more sophisticated information.

In the mid-1990's, I started connecting my computers to the Internet, thus accessing networked information for the first time since the mid-1980's when I connected to a bulletin board while enrolled in the computer course that was optional for students in my teacher education program. By the late-1990's I was teaching in a computer room full of computers connected to the Internet, and my students and I consumed information from the Internet constantly. We also created significant amount of information, but were not publishing it to the Internet.

Around the turn of the century, the old bulletin boards such as I had connected to in 1985 had emerged as discussion forums and blogs along with wikis, and other web 2.0 tools that facilitate publishing and interacting over the Internet. Soon the platforms known as social media transformed how people interact for both personal and professional purposes. Today, students discuss their work online, they share documents and edit, and they contribute to online communities that comprise local populations and global populations. In the second decade of the 21st century, users (both in schools and in the greater society) are likely to use their computers to facilitate interaction as much as they use their computers to facilitate the consumption and creation of information. In many cases, the boundaries between interaction and consuming and creating blur as a users share links to media and then contribute to the on-going discussion about the media.

Technology-mediated interaction has been the focus of intense study by various research groups, and there seems to be consensus on several conclusions about it. First, it can be used for effective for communication within dyads and small groups, and it can be used in both many-to-one and one-to-many situations. Further, digital interaction is associated with greater participation in *ad hoc* groups, which convene and disband quickly, as well as greater participation in long-lived communities. One factor associated with the extension of technology-mediated interaction for greater numbers of interactions and the expansion of technology-mediated interaction onto different types of interaction is the near-zero marginal expense of the interaction; almost all Internet-mediated interaction is free.

Further contributing to the expansion of technology-mediated interaction has been the availability of multimedia functions, which allow for more channels to be used for interaction. Increasingly, humans create digital information with their voices and with images they capture and these digital files become the focus of interaction. Students can compose and respond in audio or video as easily as they can respond with text using 21st century technology.

It is within the information and interaction ecosystem that youngsters who enter schools as students have been learning, acting, and interacting for their entire lives. Schools are charged with preparing youngsters to participate in this information and interaction ecosystem, and this ecosystem provides many functions that can be leveraged to educational purposes. Technology planning must sustain both aspects of the ecosystem.

TECHNOLOGY PLANNING

Creating a technology-rich learning environment requires the expertise of many professionals. Whereas the early history of educational technology was characterized by the "technology-savvy teacher" who assumed responsibility for managing the computers and associated devices, the time since the mid-1990's has been marked by computer and network infrastructure that requires skilled technicians whose sole function is to manage devices so that connections to the Internet are reliable. This increase in complexity has necessitated technology planning that involves multiple stakeholders.

Schools typically engage in technology planning at three levels. Teachers plan for the use of technology and information accessed via technology in the educative experiences of students. Network administrators and other technicians plan to obtain, install, and maintain infrastructure to meet both current and anticipated needs of teachers and students. School administrators plan for the financial, personnel, and professional learning needs, as well as other facilitating conditions that must be met to keep systems functioning and serving learners.

The technology infrastructure found in schools today generally comprises enterprise-level devices. Smaller campuses can generally obtain adequate access and reliability using devices marketed to small business users. The consumer products sold at retail outlets are rarely adequate to connect any but the most isolated offices in school buildings. This section describes some of the common decision-making processes and models used when making decisions regarding enterprise networks.

Decisions

Technology decision-making in educational institutions can be a complex endeavor. In large part, the complexity arises from the governance of public schools. In part, the complexity arises from the misunderstandings different stakeholders have of the work of the other stakeholders.

Elected boards of citizens are frequently responsible for decision-making in schools, and those boards hire school administrators to manage operations of the buildings and organizations; school administrators hire technology experts to manage the technology systems necessary for the business and education needs of the schools. The result is that those with the greatest expertise in managing technology systems may be relatively far removed from the ultimate decisions.

It is also true that the technology experts hired in schools often come from organizations other than education. In many businesses and organizations, the information and interaction needs of users are very specific and predictable, so systems can be designed to provide access to specific functions and information. This is a common situation in many business offices in schools. An individual may need access to certain databases or applications, and that user may never need to access other resources. In classrooms, systems must be configured for more unpredictable and flexible patterns of use, however. The differences in classroom use compared to business office use can be illustrated in two examples: operating system updates and printing demands.

Security and reliable functionality necessitates computer operating systems be updated; the process can take many minutes to complete, and computers are rendered non-functioning during the updates. In many business settings, the update has no noticeable effect on productivity; an office worker can check voice mail, have a conversation with a colleague, or catch up on other paperwork while waiting for the update to be completed. In a school, however, the same update can be very disruptive if it occurs just as students begin using computers. A group of students scheduled to use computers for 45 minutes will find a 10-minute update to be a very significant disruption in their productivity.

In most business offices, printing is a constant low-demand network need, with jobs arriving at the printer at a seemingly random pattern throughout the workday. In a computer room in a high school, however, there may be no printing jobs sent to the printer for 60 minutes, as students finish composing and editing papers. As the end of the scheduled time nears, and papers are due, then 25 jobs may be sent to the printer in less than a minute. The same system that can manage to efficiently print 25 10-page jobs sent over the course of 60 minutes may be overwhelmed by the same jobs in a short time; even if the printer manages to complete the jobs, it is unlikely to finish before the students leave for the next scheduled class to begin.

These two cases illustrate how technology experts must adapt their system designs to accommodate the usage pattern common in schools; patterns they misunderstand as they have never observed or experienced as working technology professionals in other organizations. An equally problematic misunderstanding that is the "technology-savvy" principal (it is usually a principal) who has set up his own home network, and perhaps several other small networks using consumer devices. This individual does not appreciate the differences between installing and managing enterprise networks and small home networks.

I once observed a faculty meeting that was convened to plan for upcoming network changes. The principal and the network administrator were proposing changes to add capacity and functions. The intent was to provide virtual desktops so that documents and programs could be accessed from any computer, as well as installing sophisticated network-based services including multiple printing and scanning options. A teacher captured the opinion that seemed to resonate with all of her colleagues, "I don't care what the network lets me do, as long as I know what we can do and it works every time. Otherwise, I won't trust it enough to plan my lessons around it." Many educators share that opinion.

As they collaborate to make decisions about what technology to install and how to manage it, school and technology leaders must share understanding of three ideas. First, the systems must be sufficiently secure to remain functional and reliable, but open enough to allow for the functions educators deem necessary. Second, to accomplish secure yet flexible systems, educators and technicians must engage in an on-going process to improve technology systems. Third, all stakeholders must recognize the complex nature of the enterprise networks in schools.

Secure to Open Continuum

Technology systems are very valuable. Even a modest system can represent an investment of tens of thousands of dollars for network devices (routers, switches, access points, servers, and similar devices that users never see). The cost of software to keep the devices functioning is frequently thousands of dollars per year as well. Including personnel and other expenses, the total cost of owning and operating an enterprise level information technology network exceeds the cost of the devices by several factors.

In schools, the network infrastructure is generally considered mission-critical as well; without the network critical aspects of the logistics and strategic goals cannot be met. To protect the investment and to maintain essential services, the network and information stored on it must be secured against physical threats (such as fire, flood, and theft) as well as network threats (such as viruses and other malware) and misconfiguration (through malicious intent or through accident). Securing systems against network threats and misconfiguration requires system administrators take steps to prevent unwanted changes being made to the system.

Network threats are detected and removed with (very expensive) software. In recent years, many network administrators have moved to unified threat management (UTM) devices to protect networks. All network traffic flows through this single device and software installed on it is configured to detect and quarantine or remove all varieties of malware and to prevent network intrusion. A UTM device can help identify computers that are infected with malware on a network as well as replace firewalls, content filters, and other devices that mange network traffic. Unfortunately, this software is not completely effective, but the responsible network administrator will install and use such tools to reduce threats to school networks.

The most effective step to prevent accidental or malicious misconfiguration of a system is to create and secure accounts that have administrative rights to computers, especially servers and other network devices. Only individuals who have knowledge of the network configuration should be provided with administrator accounts, and responsible network administrators will have and use standard user accounts and log on with administrator credentials only when necessary. Without an administrator account, only limited changes can be made to a computer.

If steps have not been taken to secure a computer or system, then it is open and users can make changes to the system. Potential changes include installing software (useful applications and extensions as well as malware), changing network configurations, and installing printers and the peripheral devices. While this does allow for systems to respond to users' needs and new software immediately, it does expose systems to threats from malware, or changes that interfere with the functionality of the systems.

Computer systems are not either open or secure. Every system can be placed on a continuum from open to secure (see figure 5.2). Within a school, different systems serve different needs, which necessitates technicians configure them at different places along the continuum. In general, the more secure a system is, the more reliable it is. Servers and network devices are highly secured, usually kept in inconspicuous places behind locked doors. Further logging on requires an administrator account, so students and teachers cannot access the systems with their credentials. Computers in public places such as computer rooms and libraries tend to be quite secure as well; this minimizes the potential for quick degradation.

Computers on teachers' desks tend to be the most open in a school. Ostensibly this is done to allow teachers the flexibility of installing software and extensions to make their machines flexible so they can support teaching. While this does allow teachers to test functionality without the need to seek the assistance of a technician, it is well known among technology leaders that teacher's computers are among the least functional in schools. A teacher who indiscriminately installs software and web browser extensions often will introduce conflicting software and changes that interferes with performance.

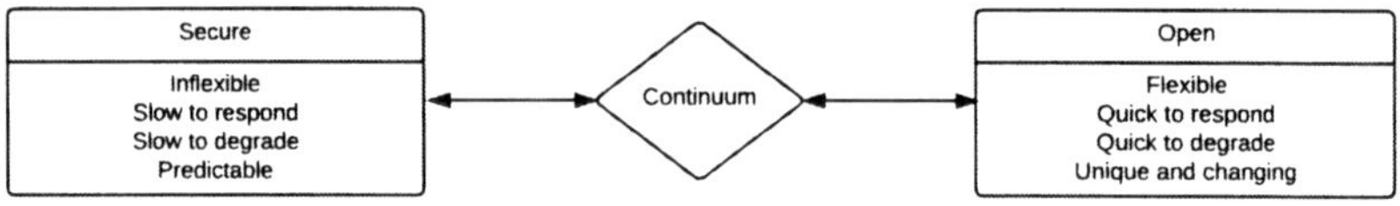

Figure 5.2. Continuum describing ICT systems

School and technology leaders must negotiate the ratio of secure to open on various systems in the school. For ease of management, and to minimize complaints of malfunctioning machines, most technology leaders argue for greater security to reduce the workload of managing devices. In some cases, computers can be set up so that any changes made can be removed when the computer is restarted. Teachers generally appreciate that feature in computer rooms and on other shared devices but not on the machines they use for their own productivity.

When planning for secure versus open systems, the question, "How do we respond to a teacher wants to install software she learns about at a conference?" often illuminates the different priorities of different stakeholders. School technology systems must allow for educators to explore new tools while encountering few obstacles and provide for rapid deployment of or access to good resources quickly all while securing data and systems within the limits of time and budget.

Planning Cycle

One of the perennial complaints of educators is that information technology systems they use at school are not configured for easy use; one of the perennial complaints of technicians is that educators do not use the systems as designed. In my experience as both an educator using systems designed by others and a technician designing systems for educators to use, both complaints are valid.

In schools in which the information technology is most effective (it serves the needs of educators and their students in a secure and reliable manner), there tends to be an on-going discourse relative to the system. This discourse is based on the assumption that the purpose of the school is to ensure that students have meaningful experiences and that decisions within the school (including technology decisions) must meet the needs of students. The on-going discourse is characterized by the planning cycle as illustrated in figure 5.3.

The cycle describes the types of interaction that are common as technology systems are designed for and deployed in schools. Generally, the stages of the cycle are clearly bounded, although different projects may be at different points in the cycle simultaneously. In many cases, also, the cycle is at different points for different individuals and on different projects simultaneously. Further, some stages of the cycle require multiple iterations before it is completed to the point that planning and deployment can continue.

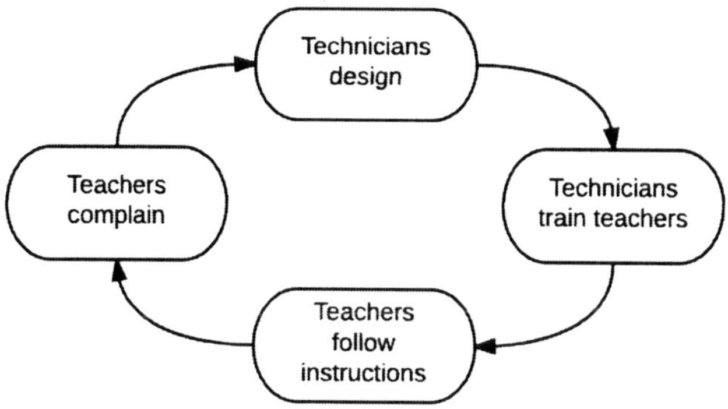

Figure 5.3. Technology planning cycle

The best time to see the cycle complete all stages in presented order with a large number of participants is the beginning of the school year. It is not unusual for summer to find system administrators and technicians take advantage of decreased summer demand and to implement system replacements and redesign plans. In the first few months of the school year, the cycle proceeds:

- Technicians plan, design, and build systems according to their skills and knowledge and as the budget and extant technology limit them. These design plans are also limited by formal and informal acceptable use policies in place in the school as well as the Children's Internet Protection Act and Child Online Protection Act and similar laws.
- Once the system is designed and built, the technicians have an obligation to ensure the teachers and other educators (and their students) understand how to use the system. This includes formal and informal training sessions, and maintaining help desks and frequently asked questions or similar self-help resources.
- The teachers then have an obligation to use the system as it has been designed and explained to them, and to insist on clarification for steps they do not understand. At this point in the cycle, educators are obliged to us the system as it was designed and explained to them. If teachers are not

following instructions, then they cannot complain systems are not functioning.

- Teachers then have the obligation to complain about the system; complain here is used only loosely. Their criticism and complaints must be in terms of ease of use and usefulness (see the section on technology acceptance on pages 114–116). Any features of the system or details of configuration that interfere with the educational use of the technology must be brought to the technicians' attention.
- The cycle continues and technicians are obligated to redesign the system to accommodate the reasonable needs of users.

When properly and fully implemented, this planning cycle results in a technology infrastructure that is perceived as continuously improving by all users. In addition, the system is adaptable, incorporating new technologies, evolving to meet new curriculum demands, and usable by increasingly sophisticated users. Successful implementation of the cycle requires deferring to different stakeholders at different parts of the cycle.

I once coached a leader through the various parts of the cycle, and the situation illustrates how the cycle can be successfully implemented. An elementary school was to receive "hand-me-down" computers from the local high school. (We will ignore the potential discussion of younger students receiving lesser technology, the reality was the teachers in the elementary school were happy to be having the machines and they would serve several educational needs the teachers had identified.)

During the summer, the technicians reformatted the computers, updated the operating system, and connected them to the network. As part of the work, they replaced the generic user name that all students in the elementary school had used to log on to the computers. Individual student accounts were configured so they would create their own passwords when they first logged on (a very reasonable step to secure a network). At this point in the story, the first step of the cycle had been completed.

When teachers first arrived in the computer room with students, they directed students to use the generic username and password that had given students access to the computers at the end of the previous school year. As that account had been disabled, students could not access the network. The technicians had failed in the second step of the cycle, which is to ensure faculty understands how to use the system; they had neglected to tell teachers, or even the principal that the log on procedure had changed.

Once teachers were given the procedure for logging on, they led their students to the step in the process where they had to create their own passwords. The default setting had not been changed on the server, so students were required to create complex passwords. While this is also a good step for securing a network, it was time-consuming and frustrating for students who were just learning the keyboard.

After two weeks of struggling to help her second grade class get logged on to the computers before their 45-minute session ended, the teacher came to me and said, "You have to help us fix this, my kids don't want to use computer because they can't log on in the time we have." (The teacher was following the procedures as designed by the technicians, thus completing the third stage.) She told me that she had asked the technicians about the problems, but she had been dismissed (I expect by a technician who was underpaid, and overworked, and trying to fix several problems that had been brought to his attention when he arrived in the building to fix something else). She had also approached the principal about the problem, but she deferred to the decision of the technology staff.

I convinced the principal to observe students in the computer room; she was quickly convinced that the log on process was interfering with the experience of the students, but she felt unprepared to propose a solution. I explained to her the conflicting problems of protecting the system with robust account credentials and providing a system that is easy to use. I further convinced her that this was a technology problem interfering with education, so the faculty must win and the students be allowed to use less complex and less secure passwords than the server was configured to allow. The network was reconfigured to allow for easier passwords, and students were able to use the computers with far less frustration, and the network reminded secure despite the less secure accounts.

Cheap, Good, Fast: Choose Two

Conflicting goals or purposes is a theme commonly encountered in technology planning. There is a well-established heuristic that originated in project management that is used by technology leaders to describe computer and network system design and purchase options for the organizational leaders. It is frequently with humor that technology leaders will say, "Cheap, good, fast, you can pick two." According to the heuristic, there are three relevant dimensions that influence the installation technology systems: cost, quality, and completion time (see figure 5.4). Clearly, leaders hope for inexpensive systems that function well and that can be installed quickly; such systems are not possible according to the heuristic and the heuristic is validated by experience.

The choices allowed by the "choose two" heuristic are represented in the circles of figure 5.4. All decisions related to information technology systems must be characterized by all three of the adjectives in one of the circles. Increasingly, school administrators appear to be taking greater steps to understand the technology in their schools than their predecessors did even a few years ago. It is not uncommon to hear a school leader actively participate is discussions related to technology and to summarize his or her understanding and seek confirmation it is accurate. Previously, it was most common to hear school leaders defer to technology leaders on all decisions related to installation and even use.

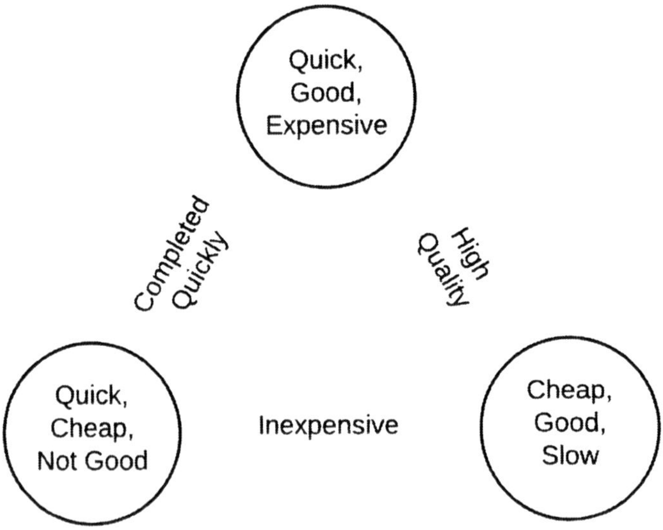

Figure 5.4. Technology decisions must fall in one of the circles.

Technology Acceptance Model

Fred Davis, an assistant professor of business administration at the University of Michigan, first elucidated the technology acceptance model (TAM) in 1989. In proposing the theory, Davis observed that information and computer technology "offers the potential for substantially improving white collar performance," but that "gains are often obstructed by users' unwillingness to accept and use available systems" (Davis 1989, 319). TAM was proposed as a tool for accurately measuring those factors that explained and predicted users' decision to use technology for a particular purpose.

In its original form, TAM posited that users are more likely to use technology for a task if they a) find it easy to use, b) find it useful, and c) believe others are using the technology for the same task. Steps taken to increase any of those perceptions among users increased the use of technology. Since Davis first validated it, TAM has been applied to a wide range of information technologies in a wide range of populations. Using TAM to measure users' experiences, engineers have developed and refined hardware, software, and interfaces that have expanded and extended technology use in many endeavors. In addition, managers have refined logistic practices so that information technology systems are used more efficaciously in diverse organizations.

By 2003, technology researchers were using eight different models to understand factors related to technology use, including TAM. Fred Davis was part of another research group that consolidated those models into one. According to the unified theory of acceptance and use of technology (UTAUT), four factors are directly associated with users' acceptance of and use of technology: performance expectancy, effort expectancy, social influences, and facilitating conditions (see figure 5.5). In addition, gender, age, levels of experience with digital devices, and voluntariness of use are identified as factors exerting indirect influences on technology acceptance In general, males are more accepting of technology than females. Young people are more accepting than older people. Those with more experience are more accepting of technology than those with less experience as are those who use technology voluntarily versus those who are compelled to use technology.

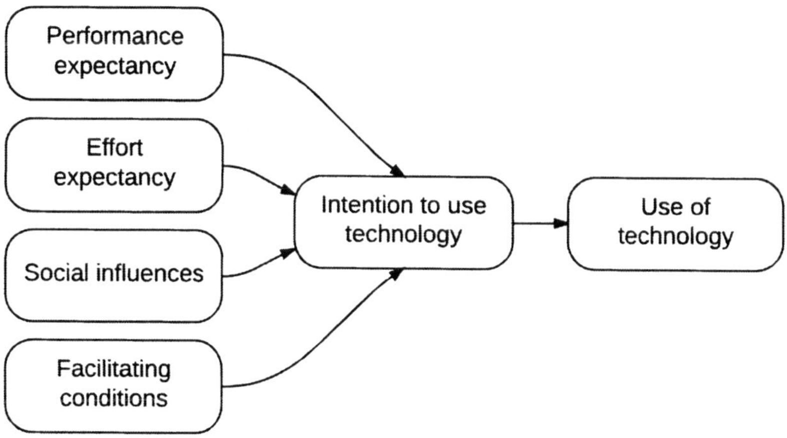

Figure 5.5. Factors directly associated with technology use (adapted from Venkatesh, Morris, Davis and Davis 2003)

Performance expectancy (PE) is a measure of the extent to which an individual believes information technology will affect his or her job performance. PE comprises two root constructs: relative advantage and outcome expectations. Therefore, school and technology leaders should expect to measure higher levels of PE in educators who perceive technology-based methods to be better or more efficient than other methods. This measure is going to be affected by the individual's perception of what performance is and how it is measured. An educator who perceives completion of instructionist-style tasks to be valuable will perceive performance differently from one who perceives performances for authentic audiences to be valuable.

Effort Expectancy (EE) is a measure the individual's perception of how easy it is to use computer technology. This factor appears to exert strong influences on decisions in the periods after initial training has occurred, but before the user becomes highly familiar with and skilled using the system. When first learning to use a technology system when they are overcoming the initial "activation energy" of learning a system, the cognitive load of using it exceeds the cognitive advantage of using a new system. When users are very familiar with the system, the cognitive load of using it, thus the effort expectancy increases so that users perceive the necessary effort to be minimal. Between these extremes, however, users who perceive the system decreases the effort necessary to accomplish a task demonstrate greater acceptance of the technology.

Social influences (SI) is the construct related to the individual's perceptions of how others perceive the technology and its use. It has complex roots emerging from the individual's sense that others (who are judged to be important) expect the tools to be used, as well as the individual's cultural experiences and expectations, and the social status that can be gained by using (or the status to be lost by the failure to use) the technology. For educators, the social influences are particularly influential, and they vary depending on the stakeholder who is experiencing the influences. For example, a new educator may be influenced to use technology in the classroom based on his or her cultural experiences, but influenced to not use it because of more experienced colleagues who dismiss the role of information technology in the classroom.

Facilitating conditions (FC) are a range of technical and organizational factors that contribute to an individual feeling prepared to use the technology systems and to feel supported in his or her use of the systems. Among the important facilitating conditions are adequate financial support to keep equipment in good repair, adequate access to prompt support to troubleshoot and repair equipment, and support for educators in their efforts to learn to use and teach with technology. FC extends as well to users' individual perceptions of technology and his or her affect towards technology. Users who perceive they have control over the computers they use, those who find the systems they use to be flexible and those with greater levels of self-efficacy tend to report greater level of FC contributing to increased use of technology (Kirschner, Pas, and Kirschner 2009; Workman 2008).

Scholars have used technology acceptance, as it was first proposed and as it was revised, to study human-computer interaction and other aspects of the technology ecosystem. In addition, technology planners have used the concept when adopting and adapting technology systems for a broad range of businesses and organizations. Both quantitative tools (surveys being widely used) and qualitative methods (interviews being widely used) have provided data to researchers and planners who seek understand technology use. Educational researchers have been slow to adopt the model, but instruments to measure technology acceptance in K-12 populations have been validated (Teo and Noyes 2008) and it appears the same factors associated with increased technology acceptance in general populations are associated with increased technology acceptance in K-12 classroom populations.

INFORMATION AND COMPUTER TECHNOLOGY SYSTEMS

For the first two decades of the history of personal computers in schools (from about 1980 until about 2000), the primary goal of school and technology leaders was to put computers on educators' and students' desks. During that period, educators were installing the school's first computers, then replacing those with multimedia-capable computers, and then replacing those with computers configured for Internet connections.

As a result, schools entered a cycle of technology obsolescence. They would expend significant capital in one year and a wave of new devices would arrive on campus. In the following years, the students and teachers would enjoy improved access to and improved function of the devices. Over time (usually less than five years), the devices became worn, increasingly dysfunctional, and obsolete compared to the products introduced to the market since the initial purchase. Recognizing the inadequacy of the systems that were sufficient just a few years previously, school and technology leaders would recommend another significant capital expense to obtain large numbers of devices that would soon be obsolete.

Since 2000 or so, the cyclic nature of technology purchases and the accompanying cycle of obsolescence have become less common. Several factors have led to tendency to more stable and predictable technology in schools. First, extant technology provides a foundation for technology purchases, and most schools have adopted a replacement cycle, so a fraction of the technology is replaced each year. Second, as networks became more important to computing, technology planning refocused from user devices to network infrastructure. eRate, a program that provides on-going financial support for network access and devices, has moderated those expenses. Third, as cloud computing has become more widely accepted, Internet-only devices have replaced computers with require full operating systems. These devices are generally priced at a fraction of the cost of full computers and the total cost of ownership tends to be less than full computers. Those factors are dampening the purchase and obsolescence cycles and they are anticipated to influence decisions related to networks and devices purchased and maintained by school technology leaders into the future.

Networks and Wireless

In 21st century schools, the data network is as important to the building infrastructure as the electric circuits. In the middle of the first decade of the 21st century, statistics from the National Center for Education Statistics (Snyder, Dillow, and Hoffman 2008) indicated that essentially all of the classrooms in the United States had access to the Internet, usually through a local area network.

Ethernet is the most commonly installed network technology in schools, and most schools have been upgrading devices to comply with Gigabit Ethernet protocols, so the networks transmit one billion bits each second.

The computer networks commonly found in schools (see figure 5.6) are probably best understood by tracing the connection from one's desktop or laptop out of the building to the Internet. By understanding the organization of the networks and the basic functions of the devices, educators can better understand the environment in which they work and provide better descriptions of malfunctioning devices to technicians.

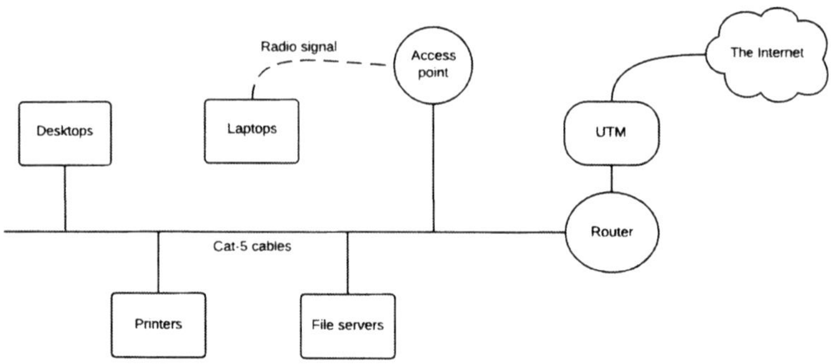

Figure 5.6. A typical school network

A computer will connect to a network via a cable that is a little smaller in diameter than a pencil. The cable (either a Category 5 or Category 6 usually called "cat-5" or "cat-6" by technicians) has a plastic plug on each end; one snaps into the network port on the computer and the other into a similar port in the wall (or sometimes a switch with many ports or a metal column dropping cables from the ceiling). A light emitting diode (LED) on the computer will indicate the cable and connection to the network is functioning; a green light on the network interface card on the computer indicates the computer is properly connected to the network.

Connecting to an Ethernet network via wireless (wifi) is slightly more complicated. A radio signal replaces the cable, and each wireless device connects to a nearby access point. The access point is connected to the network via an Ethernet cable and it broadcasts a service set identification (SSID). Modern operating systems on laptops, tablets, and smart phones will recognize SSID's that are being broadcast in an area, and will show users the names and allow users to attempt to connect to the SSID. Some SSID's require a key to connect, whereas others are public and allow any device to connect. In schools, it is not unusual for a single access point to broadcast several SSID's. Network administrators can limit networks resources, including bandwidth, which can be accessed from each SSID.

To complete the connection to the network, the operating system on the desktop computer or mobile computer will request an Internet protocol (IP) address from the device on the network that has been configured to assign the addresses. Typically a server or a router will assign addresses via dynamic host configuration protocol (DHCP); if more than one device has the same address, then neither will be able to send or receive network information until the conflict is resolved. For reasons that are unimportant to the current purposes, an IP address is given as a series of four numbers between one and 255 that are separated by dots; for example, your desktop may be at 192.168.1.184 and the network printer at 192.168.1.250. One of the first troubleshooting steps for network technicians is to "ping the IP"" in which the technician has a computer send a message to the IP address of the malfunctioning device. If the ping returns a successful response, then one knows the cables and devices are functioning and interruptions to the network service can be attributed to the devices on either end of the connection. If the ping is lost, then interruptions can be attributed to some aspect of the network.

Every device on the Internet, including web sites, is associated with an IP address. So that people can refer to network devices and locations by meaningful names rather than by IP addresses (it is far easier to remember a name than to remember random numbers), computers are assigned a domain name server to use. This server translates domain names and device names into IP addresses. If a device cannot access an Internet site by name, but can access it by IP address, then there is a problem with how the device is accessing DNS services.

On most networks, there is also a server that authenticates users. By providing a username and password, an individual verifies the identity of the person controlling the computer at an IP address, and local network resources will be available to the user at that computer. For example, schools often limit access to color printers because of the cost of toner or ink. When an art teacher logs on to a computer and sends a document to a color printer, the network will recognize she is one with permission to print in color and compete the request, whereas a student will have the request denied so the document will not print. Similar permissions are used to access file servers. Typically, a user will be assigned a folder on the server to use for saving documents. Whenever a user logs on to a computer in the local area network, that folder can be accessed; because permissions are specific to the user rather than the device, most network resources can accessed from any device.

Every network also has a router; this device keeps track of network addresses and traffic and sends it to the correct address. Interestingly, Ethernet networks allow only a single device to be using the network at one time. As a result, a message (such as a document going to a printer) will be divided into packets, and each will include addressing information. A document being sent to the printer at 192.168.1.250 will be broken into small packets, which are routed to the printer's IP address individually. Software on the printer will rebuild the packets and send instructions to the printing mechanisms.

As has been suggested earlier, network administrators and engineers have begun installing unified threat management (UTM) devices that provide services previously provided by separate devices. UTM devices are replacing firewalls. Firewall software limits both incoming and outgoing network traffic. A firewall can be configured to prevent many varieties of network traffic. Internet traffic comprises many protocols and ports, some are reserved for specific purposes, so network administrators block certain Internet services by blocking the protocol or port on the firewall. Internet gaming devices typically use specific ports to connect to game servers; by blocking those ports on the firewall, network administrators can prevent devices from accessing game servers over school networks.

On school networks, an additional device (or perhaps the UTM) will filter the Internet that sites that can be accessed. The filtering software prevents users of school networks from accessing certain content. Content filters can be applied based on category, keywords, or several other criteria. Ostensibly these devices are installed to prevent students from accessing inappropriate content, but most technology leaders also recognize the value in protecting devices from malware and minimizing access to sites that use significant amounts of bandwidth.

Bandwidth refers to the capacity of the final connection between the local areas network in the school building and the Internet. Access is purchased from a telecommunications utility and the amount of bandwidth that is available depends on the circuits maintained by the utility providing the service and the budget available to purchase the service. Bandwidth can be accurately envisioned as a pipe; the size of the pipe determines how much water flows through it and it is a zero-sum quantity. The bandwidth used for one purpose cannot be used for another purpose, which motivates many school and technology leaders to allocate bandwidth carefully.

Mobile Devices

The trend to smaller and smaller devices is well known in the history of computing. Since the middle of the first decade of the 21st century, smartphones have become a dominant device for computing. Larry Rosen (2010), a psychologist from California State University, Dominguez Hills, applied the acronym WMD to describe wireless mobile devices which he observed have become the computing device of choice for the first digital generation, and that choice was driven by the social interactions available via the devices. With these devices individuals are always connected and the devices are becoming more affordable, therefore ubiquitous. Rosen also points to two types of interaction that motivate young people to quickly adopt WMD's: connections to friends and connections to communities that help them through the difficulties of adolescence; both are socially important and motivate young people to be heavy users.

According to Rosen, users of WMD's are redefining information and interaction and their experiences are influencing their expectations of education.

Users of WMD's assume information is available everywhere and they value being skilled at finding information rather than being able to recall information. They have access to cognitive tools such as calculators and they value being facile with those tools over skill using traditional algorithms. They expect information to be pushed to them by the sources they choose and they expect to pull information from other sources as they need it and just at the moment when they need it. These characteristics of WMD-mediated information and interactions are expected to exert disruptive effects on instruction that follows the 20th century recitation script.

As they offer ubiquitous access to the vast information on the Internet and diverse systems for interaction, WMD's are challenging many of the accepted norms of technology use and management in classrooms. Students use the devices for informal learning and their brains are adapted to "reach for the phone" when they need information or interaction. These habits have become natural activities for members of these generations. School and technology leaders have decidedly less control over the information and interaction available through WMD's as well. Students do control which apps are installed on the devices they carry into schools and they can gain access through their cellular networks, thus bypassing the controls of content filters and similar systems on school-controlled networks. The devices can pose problems for school leaders who seek to control technology-mediated interaction by students.

Clouds

"Cloud computing" is the vernacular term for computing services that are provided via a World Wide Web interface. As mobile devices have become more popular, cloud computing has become popular as well. Despite the impressive computing capacity that is available in mobile devices, they have less capacity than a laptop or desktop computer with a full operating system. By transferring storage and some processing to servers on the Internet, cloud computing reduces the demand for local resources.

In his 2008 book *The Big Switch*, science writer Nicholas Carr detailed similarities between the adoption of cloud computing and the centralization of electricity generation in the early 1900's. When electricity first was used to run machines, Carr observed, each factory installed and managed its own generator. This model of on-site electricity generation was replaced with large electricity generation stations and the distribution grid that has been common for decades. (Interestingly, that model is increasingly being challenged as small-scale efforts to use renewable alternatives to fossil fuels are being developed. Those small-scale producers usually are connected to the distribution grid, however.) Cloud computing does follow the model of large scale and centralized stations that provide large capacity, and users access that capacity as needed; this is similar to how people access other utilities.

Google Drive an example of cloud computing that is widely used in education. It provides several advantages for educational communities. A user connects to Google Drive via a web browser and logs on with a user name and password, and then has access to several tools including word processor, spreadsheet and presentation software. Files created and stored in Drive can be accessed from any computer on the Internet, so the files are compatible with any computer (or mobile device) and it is not necessary to physically transport files to different locations. In addition to storing the documents created with Drive, users can upload and store files from local computers to Drive. Because the files are stored on Internet servers, they can be accessed from anywhere and they can be shared in many ways. Users can make files public so anyone can see them, or they can give others permission to comment on or even edit documents.

In schools where cloud computing has been adopted, there are several changes in technology management. More resources tend to be used to install and maintain robust and secure connections to the Internet and to ensure there is adequate bandwidth. Fewer resources tend to be spent on user devices. Because cloud computing requires only an updated web browser, devices with more limited capacity can be deployed. Chromebooks are becoming more popular as are versions of Linux with system requirements that are compatible with older computers. These decisions are contributing the dampening of the cyclic purchase and obsolescence pattern described previously. Devices that provide sufficient capacity are less expensive than other computer and devices can be used longer in schools where cloud computing is used.

For all users of cloud computing, the systems are updated and protected by the companies that provide the service. Those providers operate on a scale that far exceeds the scale of even the largest enterprise systems in schools, and thus have the resources and expertise to maintain systems with greater security and reliability than local network administrators.

Some school and technology leaders have interpreted Family Educational Rights and Privacy Act (FERPA) requirements for privacy to mean data about students that must be protected cannot be stored on cloud systems. This has resulted in two systems being maintained, one system of servers to store and manage potentially sensitive data and cloud-based systems to be used for teaching and learning.

Full Computers

The trend to smaller personal computing devices is well established. Smart phones and tablets are becoming the devices of choice for many users, and schools have been adding those devices as well. The demand for full computers, (one with an operating system such as Windows or Macintosh OS) with full sized keyboards and much more computing power than mobile devices is still considerable in schools. Mobile devices and cloud computing do provide for many educationally useful activities, but there are some that depend on full computers.

The computing power of full computers allows for applications that provide more features and more functions, and that can manage greater amounts of data than can be handled on mobile devices. While mobile devices may be useful for capturing images, video, and sound for multimedia projects, many users find the basic multimedia authoring available on apps to be insufficient for creating polished and projects that have a professional appearance. As a result, they will capture with mobile devices, but they edit projects, and render final versions on full computers.

Open Resources

A wide range of expenses consumes educational technology budgets. Very obvious expenses marked the cyclic spending that characterized the first decades of educational computing; computers arrived in boxed, as did software on disks to be inserted into the computers. As networks were installed, expenses became more hidden from users. Servers, routers, and switches (each with a price tag comparable to multiple desktop computers) tend to be installed in closets and back rooms. Those devices all require expensive software that is maintained by subscription; these recurring costs allow the systems to function, but add to the cost of owning the device. With the arrival of networks, software and other services were accessed via networks, so software downloads and license keys replaced disks for installations.

Coinciding with the transitions described above has been the increasing availability of open source software. An individual or a group will write open source programs, then release it, and allow others to make copies and distribute it without paying license fees. The open source community is highly collaborative, and the person who writes the original program will often seek the active participation of others to improve the performance of and add new features to the software.

While technical support is usually not available for open source software (one way programmers make a living in the open source community is by selling support services), there is usually a large and active community that poses questions and problems. The solutions are then made available on the Internet, thus completing the open nature of the community of users. For many purposes, open source software provides functionality equal to (or better than) that available from expensive proprietary software.

In many schools, open source software has been used to minimize demands on technology budgets. Ubuntu, a version a Linux, has been developed as an alternative to Windows, and it is not uncommon to find it in schools. In addition, many pieces of open source software for managing data and networks are available. Most technology professionals are comfortable with the nuances of using this software that can decrease the ease of use for less experienced users.

VIRTUAL CLASSROOMS

As the World Wide Web matured, the capacity of the web to support educationally relevant tasks developed as well. Since early in the century, easy web publishing via blogs, along with social bookmarking tools, file-sharing services, chat rooms, forums, and other tools for interaction and information have been available for educators to use. More recently, web-based tools for creating portfolios, administering tests, online rubrics templates, and even cloud-based grade books have become available. While these tools were useful, each tool tends to provide a single service. As a result, many educators find that using different systems for each function poses a serious obstacle to using them. Further, most of these sites require users to create accounts, and the terms of service prevent those younger than 13 years from creating accounts.

To make the many educationally useful services available through a single source that is managed by local school and technology leaders, schools can obtain a learning management system (LMS). This web-based system incorporates many of the useful web 2.0 tools and functions into a single platform. Through the LMS, all teachers in a school can create a virtual classroom and students enroll in the courses, thus accessing the virtual classrooms their teachers create.

For an increasing number of educators, an LMS is as essential as any other computing infrastructure. For these teachers, their classrooms are both physical places and online spaces. In these hybrid classrooms, teaching and learning occurs both in-person and online and each type of experience complements the other. Taken to the extreme, an LMS can be used for online learning and entire schools exist in which students and faculty meet and interact only through web interfaces. Virtual classrooms are anticipated to play an increasingly important role in K-12 education as more technology is leveraged for more educational tasks.

Functions

The tools provided by a virtual classroom are designed to reflect traditional classrooms. Virtual classrooms accommodate all aspects of the technology ecosystem, and the system can be configured to accommodate local needs. An LMS can be connected to a student information system (SIS) so that courses, enrolments, and grades can be managed in one and synchronized to the other. This usually requires scripting that makes the LMS and SIS databases compatible, which requires significant work and usually involves multiple programmers who are deeply familiar with the LMS and SIS.

Teachers who have adopted web tools that provide individual functions are sometimes reluctant to stop using their familiar tools, and to begin using an LMS. They recognize (accurately) that the switch will necessitate effort on their part. In general, however, the advantages of having all of the tools available through one interface that is common to all users in the school and that is controlled by the school are strong reasons for making the switch. In addition, many teachers use web-based tools that are driven by advertisements, so they are requiring their students be exposed to commercial messages as part of their school experience.

Content Management

Just as in-person schools are organized around courses and sections, an LMS is organized around courses and sections. Creating a course on an LMS is a task performed by the LMS administrator, unless the LMS administrator has configured the system to allow teachers to create their own courses. Once a course has been created, there are many options for creating content in it.

Most LMS make template functions available; a course is created with links (perhaps to the school library), handbooks, policies, calendars, grading systems, and any other content deemed necessary, then it is saved as a template. When a course is created using that template, any contents in the template will appear in the new courses. A science department may find it useful (for example) to create a template with information regarding laboratory safety and links to online periodic tables, and similar references that are needed by students in all science classes.

Related to templates are parent and child courses. This feature is frequently used when there are multiple sections of a course taught in a school. A single parent course will be created, and students enrolled in child courses, one for each section. When a change is made to the parent course and then the parent and child courses are synchronized, any changes in the parent course are will appear in each child course. In a school that offers several sections of Algebra 1, for example, the math department may specify all students take the same tests and exams, but allow teachers to give unique quizzes to their students. A parent course containing the test and exams could be created, and those synchronized to the child courses. Quizzes in child courses would be available only to these students in those sections.

Some publishers provide teachers with supplementary resources for their textbooks available in a compressed file that can be uploaded to an LMS. Once extracted, the LMS course is filled with all of the publisher's resources. Teachers can then modify the resources to meet their specific needs. Other publishers will make content available as sharable content object reference model (SCORM) packages that can be uploaded to a SCORM-compatible LMS. The zip files and SCORM files are local resources. The file is uploaded to the LMS from a hard drive, and the files are then incorporated into the LMS. Editing the files is then done through the LMS.

Learning tool interoperability (LTI) is another method whereby the content created by on one LMS can be included in another. Using LTI requires the LMS administrator to configure the two systems to share information, and changes made to the original LTI resource will be reflected where it is embedded. LTI is different from SCORM in the location of the files. LTI tools exist on a system other than the one where the LMS is installed. Changing an LTI resource requires accessing the host system, and if the host system becomes unavailable, then the LTI resources are unavailable.

Further, an entire course can also be downloaded from an LMS. When compressed and downloaded, the course is saved as a compressed (.zip) file that can be uploaded and extracted to the LMS, thus recreating the course. Compressing files in this way can be used to export the content of a course or to archive it. While a course export will save all of the information contents of a course, a course archive will include all of the information and all do the interaction, including student submissions and grade book entries for a course. Some schools specify an archive procedure to provide evidence if disputes arise regarding a course.

Each LMS also provides a variety of content creation and editing tools. Hypertext markup language editors are available on each LMS, and teachers can embed media from other web sites into the html objects they create in their virtual courses. LMS courses can incorporate discussion boards, chats, and a wide variety of other tools for online interaction that are provided by the LMS. Further, each LMS includes options for testing and other assessments.

File Sharing and Linking

One of the functions that educators tend to use when they first begin using an LMS is the file sharing and linking functions. This feature is similar to other web-based tools for file sharing; a file can be uploaded and linked to a post, which is created with the html editor. Study guides, articles, and presentations used in class, images, and similar files can all be uploaded to the virtual classroom. Further, uploaded files can be placed in either the individual teacher's library or the school library, so files can shared with minimal effort. For most teachers, the advantage of sharing files via a virtual classroom is that they are relieved of the responsibility of replacing lost resources. Students or parents can access the course materials whenever necessary.

Although file sharing, html, and links can be managed by many others tools, the LMS also includes sophisticated tracking and management options. Teachers can see who has viewed resources and the length of time they viewed each; this feature is very valuable when used to track test performance. Resources may be hidden from students, and made available only under the conditions specified by the teacher. This tool is useful for unit development purposes, as teachers can edit resources in a space unavailable to students, and then open the spaces after the materials are finalized. In addition, teachers can open resources based on dates or performance conditions. For example, a teacher can require students repeat a test until a threshold score is reached; only after that score is reached will the next set of resources be available to a student.

Interaction

Many varieties of web 2.0 tools have been available since the late 1990's; these tools are all designed to make it easy for users to publish information to the web and to interact with others via posts and responses. Many of these are built into

LMS, so can easily be incorporated into virtual classrooms. The choice of which tool for interaction to use often depends on the purpose of the tasks and the nature of the interaction the teacher seeks to facilitate.

Threaded discussion boards are used in situations where the teacher provides a prompt and then students compose responses. Others can reply to any response, and each reply is indented under the response (or reply) to which it replies. The result is an outline-like organization with levels containing the responses to the same response or reply. Many find the interaction on threaded discussion diverges from the original prompt. In some situations that divergence is distracting and contrary to the learning goals, but in other situations, that divergence is the intended outcome. This can also lead to meaningful explorations and connection making.

Blogs are used when the teacher seeks to have students post in response to an initial prompt, and then encourage dialog that is focused on each individual's response to follow-up. With blogs, the discussion tends to focus on one individual's response, and others add comments to blog entries. Many students and teachers find the blog interface can be much easier to follow and leads to deeper discussion than the threaded discussion board. Many LMS's also provide a private blog tool that can be seen by the student and the teacher; these private blogs can be used a journals for individual learners.

Wikis are tools that allow for a group to compose and edit a file or collection of files. Often these are used if a task has been divided among individuals or groups and each is contributing a part of the whole. Wikis are also useful for idea improvement tasks. In this situation, a partial answer may be provided and individuals or groups edit the answer to make it more accurate or clearer. In many LMS systems, the instructor can lock a wiki once a "good" solution or answer is provided.

Chat provides a synchronous method of interacting via text. When chatting, individuals type a message, which is sent to the chat room where others can read and respond. When the chat contains many participants, there can be multiple connected (or disconnected) conversations on-going. Just as with threaded discussions, this can be distracting or it can be desired. Most chat platforms include an option for recording a transcript; many educators find the transcript of a chat can be used for other activities. For example, the transcript of a chat regarding the characters in a novel can be used as a source for writing profiles of the characters.

As broadband video and computers with web cameras built in have become widely available, video chat is also arriving in LMS courses. Video chat can be quite sophisticated and include capacity for many sites or users to connect. Once connected, participants typically have access to chat, and they may have options for participating by voice or even sharing their screen with the group or controlling a computer located at another site through the video chat site. Video chat does require significant bandwidth, and the services can be very expensive if one purchases access to systems designed for enterprise level video chat.

Teachers also have the option of creating groups of students within an LMS. Depending on how the groups are configured, the members can have their own collection of information and interaction tools that are available to group members by no others, and assignments can be given to students based on their group affiliation.

Assessment and Evaluation

Each LMS includes a full complement of tools for assessing and evaluating students' progress, including increasingly sophisticated tools for real-time performance tracking and flagging or warning poor performance and analyzing performance on test items.

Online tests can include questions that can be graded by the LMS, so scores are immediately available and recorded. Questions that can be scored by the LMS are multiple-choice, fill-in-the-blank, and others that can be clearly and accurately checked by comparison; accurate scoring of these items requires the teacher (or author) correctly specify which is the correct answer. These tests can also include items, such as essay questions, that must be scored by the teacher. Regardless on the question type, feedback can be left automatically by the LMS, so an incorrect answer may point students to a page number or provide a link to a resource that clarifies the idea. Teachers also have the option to compose feedback to individual students, and the teacher can overwrite test scores recorded by the LMS.

In addition, there are sophisticated options for varying time limits, numbers of attempts, and other options regarding the test presentation. Most LMS's also allow for individual exceptions to test settings. These settings can accommodate the needs of specific students or allow students multiple attempts until a certain score is achieved.

LMS's also provide drop boxes for assignments that are created using other software. For example, students in mathematics class who create an assignment in a spreadsheet can upload the workbook. Each submission is archived with a time and date stamp so there are no questions about when the file was uploaded, and the contents of the original can always be retrieved. In addition, these systems provide various options for submitting to (usually fee-based) plagiarism checking services. Further, some provide the capacity for in browser commenting so a teacher can mark-up a submission without downloading the file to his or her computer.

Most LMS's also provide rubric-writing tools as well. By composing text in a series of boxes within the interfaces, one creates a rubric. Once it is created, a rubric can be attached to assignments and used to score submissions. In addition, most LMS's allow teachers to specify learning outcomes for the course; the learning outcomes can then be attached to specific assignments. Outcomes and rubrics are often included in course templates or parent courses so that they can be deployed across many courses with no effort from the teacher.

A fully customizable grade book is also available in all LMS platforms. Online tests can be automatically added to the grade book and columns with non-LMS assignments can be added as well. LMS grade books are also configured to support feedback and interaction between teacher and student. It is unusual for the LMS to be used as a student information system (SIS); in most schools an SIS is adopted long before an LMS. Connecting the LMS and the SIS is possible, but it is beyond the scope of local technology leaders. The programmers of the LMS and the SIS must write a program that connects the databases used by each.

Platform and Installation Options

LMS platforms are available from both proprietary and open source publishers. In addition, content management systems (CMS), which are used to create web sites that make use of modules and that frequently allow for interaction, and make different contents available according to different accounts, are often described as useful for educational purposes. While CMS tend to provide the same information and interaction options as any LMS, they generally do not include the assessment and evaluation tools. Some of the platforms that provide full LMS are identified in table 7.2.

In general, the features that are provided in one platform are available in the others as well. All of the platforms rely on sophisticated web databases to manage the enormous amounts of data that comprise an LMS course. For that reason, installing an LMS requires more than casual knowledge of web hosting, and the LMS must be installed on a server that has both web server software installed and also the correct version of an Internet database installed and properly configured.

The total cost of any LMS platform depends on how the software is hosted and which features are installed. Most commercial LMS providers offer multiple levels of support and service, and features purchased from third party providers can extend the costs. When budgeting, school and technology leaders are likely to plan tens of dollars per user per year to provide a commercial LMS. An open source LMS is free to obtain, but then the school and technology leaders are responsible for managing and securing the server on which the system is hosted.

School and technology leaders can make LMS available in one of three ways. Each has advantages and disadvantages, including those associated with cost and control over student accounts.

Most publishers of LMS provide limited versions of their LMS at no cost to anyone with an email address. While these systems are free, and can be used for some purposes in K-12 communities, most school and technology leaders find there is insufficient control over accounts and content to use this option with students. In addition, the free versions tend to have limited functionality, and cannot be customized by users.

Table 7.2. Some commonly encountered LMS

Name	Description
Blackboard	a commercial enterprise that purchased many competing platforms several years ago
Moodle	an open source system that has been around for many years
Google Classroom	a limited LMS available to users of Google for Education
Canvas	a relatively new system that is marketed as a highly customizable LMS, used in many small and niche markets
Edu 2.0	a commercial LMS popular in K-12 markets
Desire2Learn	a commercial provider from Canada

Publishers of LMS or third-party providers make hosted LMS packages available to subscribers. These installations are made on servers controlled by the provider, and they take full responsibility for ensuring systems are backed-up, secure, and otherwise protected from threat. A school that purchases hosted service is given full control over the system, so school and technology leaders can manage all aspects of the system. In general, these are very reliable, and very expensive.

Either proprietary or open source LMS platforms can be installed on servers owned and controlled by the school. This option requires the greatest level of skill and the greatest level of risk as school and technology leaders are responsible for all aspects of managing and protecting the system.

Access to an LMS (whether public, hosted, or installed) is controlled by a user's account, and the accounts can be configured for varying levels of access and control. These levels of access are common:

- *Administrator* accounts can control the rules for course creation, such as who can create or request courses, how templates are managed, and connections to other LTI systems, and other aspects of the configuration. These accounts also control user accounts. These are typically reserved for technology leaders, network administrators, and school administrators.
- *Instructor* accounts can add content to courses that have been created by administrators and (usually) manage users in the course, as well as access students' contributions to the course, their tests and assignments, and the grade book.

- *Student* accounts can access the courses in which they are enrolled and access whatever resources, assignments, and grades instructors have made available.
- *Observer* accounts are usually associated with a student, and that account can observe (but not impersonate) that student. Special education teachers, parents, and other adults can be provided with observer accounts.

Especially in schools enrolling many students and with sufficient organizational structure, other roles are created to manage student data and to create and manage content in courses. In situations where the curriculum is highly prescribed, teachers may be able to interact with students, but not change the course content. In other situations, teaching assistants who can interact with students, but not observe student grades may be given specialized accounts.

Chapter 6
Developing Educator Capacity

In previous chapters, it has been demonstrated that the information landscape of the 21st century is different than it was in previous decades. It has also been established that the nature of educative experiences for the current and future digital generations is different than it was for earlier generations. Further, it has been established that the educational technology landscape has changed and is continuing to change. From this series of observations, it is reasonable to conclude the future of teaching and learning in classrooms is difficult to predict, and educators are preparing students to manage unknown problems. It is also reasonable to conclude that educators must be prepared to engage in on-going professional learning, so they can manage similar unknown problems.

In drawing parallels between the history of life and the history of societies (including our 21st century society), biologist Geerat Vermeij observed, "the most effective adaptation for dealing with unforeseen circumstances is adaptability" (2010, 81). This theme has appeared in the literature regarding teacher preparation as well. Ben Williamson (2013) used the term flexible specialist to describe the type of worker needed by innovative organizations; by extension the same term can be applied to the educators who prepare students who will become those workers. Sawyer described teachers in the future as knowledge workers who "deeply understand the theoretical principles of how students learn" and who are "deeply familiar with the authentic practices of scientists, historians, mathematicians, or literary critics (2006b, 572)." Sawyer further concluded, "The classrooms of the future will require more autonomy, more creativity, and more content knowledge (2006b, 572)."

In the dynamic world marked by rapidly evolving information technology and expectations regarding literacy, professional development for educators must be adaptive and be designed to develop capacity similar to the capacity students develop. This chapter uses the unified theory of acceptance and use of technology (which was introduced in the previous chapter) as a framework to understand systems and strategies for building capacity for adaptability in populations of educators.

EFFORT EXPECTANCY

As effort expectancy increases, users perceive the technology to be easier to use or that tasks can be completed with less effort if technology is used. If an educator perceives it takes less effort to complete a task by using technology, then the educator is more likely to use it to complete the task. Continuing to improve effort expectancy within a faculty is difficult as devices and systems are continually replaced and updated.

The history of electronic digital information technologies has been marked by development and refinement. A device new to the market with the fastest speed and greatest capacity will be made obsolete by the next generation device which will be on the market within months. This means that educators, whose careers extend more than a few years, will experience the arrival of new devices, platforms, and information sources multiple times. Each arrival, whether a school-wide replacement of many devices or the arrival of a single new laptop on teacher's desk or the addition of new features to the LMS, will necessitate the educator adopt the new technology and adapt their computing practices to the new features. For many educators, the transition from a comfortable technology to a new technology is accompanied by a (temporary) disruption in effort expectancy.

Obsolescence threatens effort expectancy in schools that were early adopters of technology. Educators working in those organizations that adopted (for example) the first digital student information systems may find that new systems are more functional and easier to use, but problems of data preservation and interoperability limit options for replacing the older and more difficult-to-use systems. Regardless of options for replacing systems, effort expectancy can be improved when school and technology leaders take steps to address issues related to basic operations of computers, when they recognize the role of cognitive load in ease of use, and when they develop strategies and organizations to help educators overcome the demands of learning to use new systems.

Operating Technology

When computers first arrived in classrooms, they were unfamiliar to many teachers. The simplest tasks of powering computers on and loading programs required training. One of the first research programs to study classrooms in which computers were available was the Apple Classrooms of Tomorrow (ACOT) project. Those researchers discovered that training teachers how to use computers was necessary, but not sufficient, for them to be used as effective learning tools (Schofield 1995).

Grounded in the lessons learned about technology and teachers from ACOT, school and technology leaders tended to develop professional development programs designed to address three related but different aspects of classroom technology: a) the basic skills of operating and using devices and information sources, b) the design of technology-based curriculum and instruction, and c) managing technology in the classroom. The extent to which each aspect occupied the attention of technology-rich professional development has changed over the decades. Basic operations of computers and managing classroom technology have both largely disappeared from professional development as technology has become more familiar and as technicians have been hired in schools. Teaching and learning with technology continues to be revised as various pedagogies have captured the attention of leaders.

In the first decade desktop computers were in schools, most teachers had little experience with computer technology, so dedicating professional development resources to train teachers in the basic operation of systems (tasks such launching applications, creating and editing documents, and saving and printing) was appropriate and necessary. Soon thereafter, the local area networks and the Internet arrived in schools, so training teachers in the basic operation of network systems, including tools like search engines and email was necessary and appropriate.

Now that networked devices have been available for decades and become mainstream as greater parts of the population became users, it seems reasonable to expect that anyone who seeks to work as a professional in education will arrive on the job prepared to operate or learn to operate desktop, laptops, and even tablet computers they (or students) are given. Further, they should arrive with the ability to adapt their skill to new models of computers, and with a general awareness of how computers function and how to create documents and navigate networks. While no list of technology skills can be complete, the following seem to be skills every educators should arrive in the classroom prepared to perform independently:

- create and manage documents with any productivity suite, including cloud platforms
- find, share, and use Internet resources with facility
- publish information to the web
- use email
- model ethical and safe computing
- perform basic troubleshooting and make informed requests for assistance

The details of how to perform these operations on the specific systems provided in each school are likely to vary, so a part of the on-boarding process of newly hired educational professionals must be to ensure they are given instruction in logging on to and accessing necessary technology systems. This is of particular importance when the systems are necessary to perform their duties; such systems include email, the student information system, and virtual classrooms. Once an educator is shown how to access systems for these tasks (for example, once they are given a username and password and a web address to get their email), the details of how to perform these tasks on specific systems should be within their ability to learn with independence. Educators can and should assume responsibility for being able to use (or quickly learn to use) new or updated operating systems, productivity software, and emerging Internet-based information sources. Updates and simple changes to the interface should cause little consternation to educators in the 21st century, and should not be an excuse for not using a system.

Cognitive Load Theory

Cognitive load theory (CLT) was developed in the 1980's and it arose from the observation that each individual has a limited amount of cognition that can be applied to a situation. Within the situation, cognition must be dedicated to both processing information and understanding the information. According to CLT, cognition is a zero-sum quantity; so the amount used for one purpose is unavailable for other purposes (van Merriënboer and Sweller 2005). For educators, who seek to adopt technologies for their classrooms, understanding cognitive load theory will allow them to explain and predict which steps can increase effort expectancy.

Sources

Cognitive load theorists recognize three types of cognitive load: *Intrinsic cognitive load* is associated with the learner thinking about the information and the task. Intrinsic load does increase as the task becomes more complex, but steps to break the task down into parts and the use graphic organizers (for example) to help store and organize information are strategies for reducing this load. *Germane cognitive load* is that used to construct new knowledge. Building new connections, creating metaphors, summarizing, and generalizing are all products that arise from germane cognitive load.

Extraneous cognitive load is that associated with poor design or with other noise in the setting. Educators take steps to minimize extraneous cognitive load in its many forms. In a simple example, more than one person speaking at a time can increase the extraneous cognitive load of a classroom, as one must use some of their zero-sum cognition to focus attention on one speaker. Those who find it difficult to learn in noisy settings are experiencing the effect of reduced germane cognitive load because of excessive extraneous cognitive load. The noise can affect all senses. When using technology, unfamiliar or complex interfaces along with controls that are difficult to find all increase extraneous load.

Figure 6.1 illustrates the changes in the distribution of load in two situations. In the two cases, we assume the task requires the same information to be held in memory, so the intrinsic load is the same. In the top case, the extraneous load is relatively small and there is a greater portion of the cognitive load available for germane load for new learning. In the bottom case, the task is being performed in a setting with greater extraneous cognitive load, thus less germane cognitive load is available for new learning. When using technology for the first time or when using poorly designed technology, extraneous cognitive load associated with operating the technology and navigating the interface increases extraneous cognitive load, so it resembles the bottom case in figure 6.1.

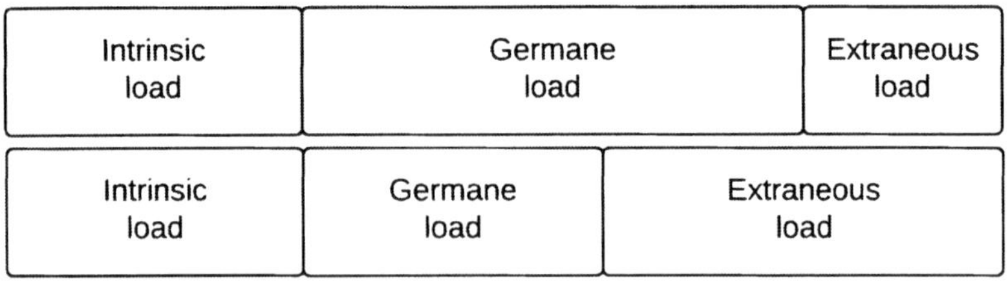

Figure 6.1. Two situations illustrating cognitive load

Managing Cognitive Load

There is a complex relationship between technology and cognitive load. In some situations, technology tools can reduce intrinsic and extraneous load; and in other cases, technology tools can increase those loads (van Merriëboer, Kirschner, and Kester 2003).

Landauer (1996) described the early process of technology design as engineers building functions into systems and interfaces for users to control the functions but the systems frequently failed to perform as expected. In many cases, the design was found to introduce excessive extraneous cognitive load. When engineers began to accommodate users' habits and needs and to include end users in the design process, the failure rate began to decrease. By focusing on the learners' perceptions and experiences with technology, educators can use the same situational awareness to improve the effort expectancy of digital devices in classrooms.

Educators have used cognitive scaffolds, such as graphic organizers for decades to reduce intrinsic load. Templates and easy-to-use drawing tools that are built into word processors are examples of technology-based tools that have been adopted for increasing the effort expectancy of using technology to support learning.

On-demand delivery of procedural information is another method of reducing cognitive load with technology. By making steps for completing an algorithm in mathematics (for example) available when the need arises, educators can reduce the extraneous load of remembering the steps, and also make the information available when it is relevant and the focus of germane cognitive load. Further, complex tasks, such as graphing, can be accomplished via technology. When the cognitive load of plotting points and drawing the curve accurately is transferred to a device, learners can use more germane load to understand the meaning of the graph.

By using the same software for similar operations in different settings, educators can reduce the extraneous cognitive load of using unfamiliar software. For example, if both the science and math teachers in a school use the same programs for creating graphs with students, then the extraneous cognitive load of using the software will decrease as the operation of it becomes more familiar. Additional germane cognitive load will be available for students to reflect on and understand the meaning of the graphs in both classes. Further, in schools where computers are used frequently, the extraneous cognitive load of using the devices is minimized, as students become very familiar with logging on and accessing necessary systems and resources.

Initial Learning

In chemistry, one learns about activation energy. This describes the amount of energy needed to begin a chemical reaction. It is well illustrated by igniting a candle. To begin, one has an unlit candle and a match; by striking the match and touching its flame to the wick, the candle begins to burn and we can enjoy the light. Without the activation energy of the friction to ignite the match, the candlelight cannot be enjoyed. A similar circumstance exists in the operation of technology and its application to tasks.

When properly applied, technology does increase the efficiency and efficacy of tasks thus increasing the effort expectancy, but learning how to perform the task with the technology requires time and effort. The effect is illustrated in figure 6.2.

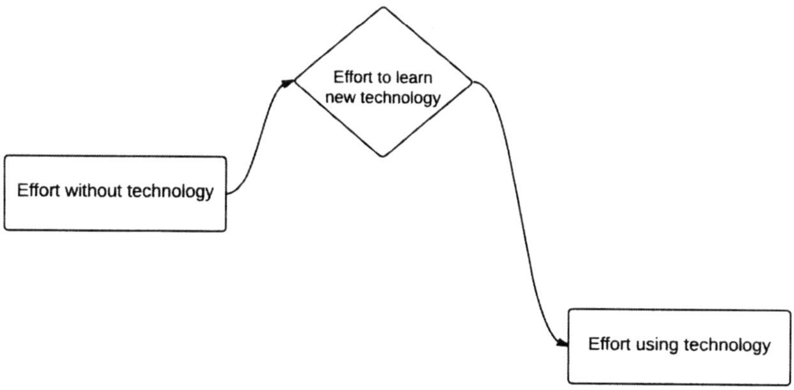

Figure 6.2. Learning new technologies requires temporary decrease in EE

Consider the example of quizzes. Without technology, the teacher composes the quiz then creates a copy for each student. Once the quiz is completed, the teacher collects each copy and reviews it to determine scores and enters the scores in a grade book. At each step of the process, time and energy is required. If the teacher sets up the quiz to be delivered using technology, then there is an investment in time and energy that is necessary to learn to use the system and then to create to questions and identify the correct answers. Once the computer-based quiz is created, however, the effort is much reduced. The quiz can be administered, graded and entered into the grade book automatically. Further, the same teacher or other teachers can reuse the automatically graded quiz with a few mouse clicks. After the initial work of creating the quiz, it is administered with far less effort.

The example may have been poorly selected as the principles of deeper, active and authentic learning minimize the importance of quizzes and similar evaluations of learning. When I taught science, however, I wanted my students to know the parts of microscopes before we used them so they could understand the directions I gave and we could avoid damage to very expensive devices. Each year, I created and copied and graded and recorded grades for this simple learning task. If my colleagues and I had electronic quizzes available, then the repetitive task of administering the quiz could have been automated and my students and I could have focused more time on using microscopes to see the unseen. In this manner, technology can improve the effort expectancy of simple assessments of learning and make time and energy available for other educationally relevant purposes.

PERFORMANCE EXPECTANCY

One is more likely to use technology if it is perceived to be useful and to improve either the efficiency of one's performance or the quality of one's performance. Central to the definition of performance is one's perceptions of the importance of the performance as well. If technology facilitates performance on tasks that are not deemed important, then improved performance expectancy is not met.

This observation was established early in the history of personal computers in classrooms. In 1997, Judith Haymore Sandholtz, Cathy Ringstaff, and David C. Dwyer identified three stages of concern regarding teachers' perceptions of computers in classrooms. When computers entered the classroom, teachers found only weak connections to the tools and their curriculum, so entry into technology-based teaching was slow. The same observation of slow entry has been made when new devices, new software, and new information sources become available to teachers. Especially as teachers first learn to use new technologies, the effort expectancy is low and that affects performance expectancy, so entry is characterized by low levels of use for multiple reasons.

Sandholtz, Ringstaff, and Dwyer found, however, that as educators become familiar with computers especially for their own purposes, they adopt it to support teaching and learning. This tends to be a self-reinforcing process; as they become more capable of using the technology and more aware of how it can improve teaching and learning, their use increases. As the usefulness of technology becomes clearer, teachers will progress from entry through adoption, and they may adapt their curriculum to both leverage and reflect the technology. Performance expectancy for individual teachers increases as they progress from entry and through adoption to adaptation.

Consider the work of researching. Today, researchers have a number of tools available for managing bibliographies. Some will allow users to add a work to the database by entering the International Standard Book Number (ISBN) of a book or the digital object identifier (DOI) of an article. These tools are connected to machine-readable databases, so the details of author, title, publisher and similar facts are added to the researcher's database without effort. These tools also allow for formatted bibliographic entries to be copied and pasted from the database into a word processor. When these tools enter a classroom, teachers commonly move quickly to adaptation and will incorporate the tools into research assignments and include instruction in using the tools. For those educators who value students identifying details of references or who value formatting bibliographies as part of the research process, tools that automate the work will not increase performance expectancy.

Performance expectancy can be increased in technology-rich teaching and learning when educators recognize reflexivity to more quickly adapt their teaching to technology. In addition, performance expectancy can be increased through on-demand instruction, and through the use of technology to manage classrooms.

Recognize Reflexivity

The term reflexive was originally used to describe the reciprocal and simultaneous effects between social science researchers and the subjects of their research. In the research focusing on the modern information technology and its role in creating the sociocultural context of the 21st century, the term reflexive is used to describe the reciprocal influences of technology devices on the tasks necessitating (and necessitated by) information technology and the individuals and groups who use that technology for those tasks (de Vaujany 2008). In technology-rich teaching and learning, the pedagogical choices are affected by the technology that is available and technology choices are affected by the pedagogy that exists or that is desired (see figure 6.3). These reflexive relationships generally lead to expanded and increased use of digital devices for a greater number of tasks or for greater and more diverse populations, but reflexive influences can lead to interactions that are qualitatively different from interactions with no technology use.

In the example of the bibliography tools used to introduce this section, we see an example of how technology is changing the nature of the work done by researchers. Especially with reference to information tasks, the nature of the tasks is changing as the result of new information technology tools. Our tools and our practices are exerting reflexive forces.

In educational organizations that recognize reflexive relationships, leaders will make technology decisions based on the devices and systems that create effective teaching and learning environments. Simultaneously, teaching and learning decisions will be made to reflect the skills needed to use available technology. The reflexive influences are real and active and emergent. Technology determines what we teach and how we teach it. What and how we teach informs technology decisions.

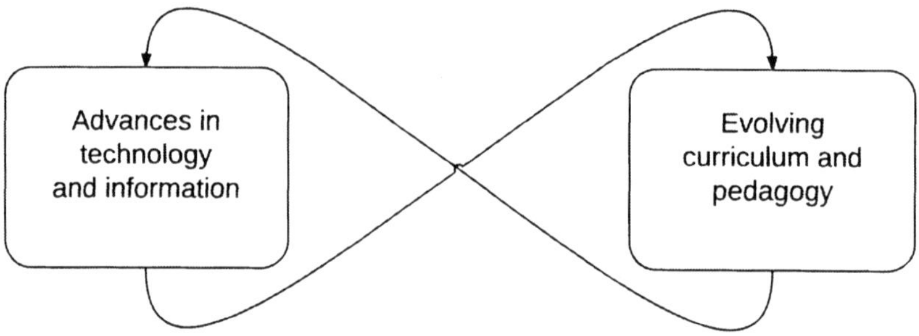

Figure 6.3. Reflexivity in the evolution of technology-rich teaching and learning

In the planning for instruction in the 20th century, technology was understood as a neutral aspect of the learning environment. Advocates of technology-rich classrooms recommended using it as a tool for instruction, but they did not recognize the influences that the technology exerts on students' experiences and expectations and their patterns of cognition. In a popular textbook for courses designed for educators learning to create technology-rich classrooms, Robyler (2006) defined integrating educational technology as "The process of determining which electronic tools and which methods of implementing them are appropriate responses to given classroom situations and problems" (9). It appears that advocates for this type of technology integration approach curriculum and instruction planning as predictable; educators proceed as if they know unambiguously and in advance which technologies will accomplish which goals. Many scholars including information technology theorists would respond that such predictability is not possible, and technology will influence the social experience of teaching and learning in an unpredictable and ambiguous manner, so Robyler's prediction is not possible.

Examples of the reflexive influences of technology on education are illustrated in the pillars of digital learning elucidated by Cathy Davidson and David Goldberg (2009) who are associated with the John D. and Catherine T. MacArthur Foundation's Digital Media and Learning initiative (see table 6.1). Davidson and Goldberg identify learners' experience with technology (especially social networks) outside of formal education, new sources of information and expertise, and flexibility as characteristics of informal learning that will increasingly influence what happens in classrooms. As these aspects of technology-rich society become more embedded in learner's experience, schools will respond reflexively. Independent and autonomous learners as well as tasks that reflect the nature of information and interaction in the digital world will increasingly characterize classrooms. Learning environments are also expected to become more flexible and creative in both the nature of the curriculum experienced and the nature of the resources used.

For 21st century school planners, this means that deep, active, and authentic curriculum and instruction, along with technology will blur into one aspect of the classroom; the information we teach and the technology we use to search and access, consume and manipulate, create and share information will become increasingly interconnected and inseparable aspects of knowledge-building. This blurring of factors explains in large part the need to apply hermeneutic methods and the non-linear design methods that are typical when working with wicked problems. Technology planners will increasingly respond to both new technology tools and to the demands of teaching and learning when designing systems. Educators will increasingly respond to new information technology tools as they design classroom experiences.

On-Demand Instruction

Especially when educators include significant amounts of deeper, active, and authentic activities in their classrooms, there is frequent need for students to be reminded of how to perform certain tasks or to solve particular problems. This need often arises at different times in different students as they study independent inquiries, therefore demand is irregular and for small groups. In these situations, the tasks become relevant in a manner they were not previously, so the situation becomes what my education professors in the mid-1980's called "teachable moments." With technology, educators can point students to instruction that is stored as video or other presentations, and provide information just when it is needed.

On-demand instruction improves performance expectancy in those purposes that are well suited to instruction. Knowledge and skill that can be broken into steps and that can be clearly assessed and evaluated (tasks and skills that are tame) can be addressed through this approach.

When working with emerging researchers to compose their first research papers in which they are expected to follow specific formats for the references page, I often will give a lecture on how to properly format references. I present a slide show that incudes a step-by-step example. After that lesson, the slide show is available on the virtual classroom, so that students can refer to it as they finalize their research papers. This presentation is also made available to other teachers in the school, so they can refer to it when they assign research papers. Typically my presentation is given to students early in their careers, and then it is accessed on an on-demand basis through virtual classrooms later.

On-demand instruction can also be used to explain complex concepts to learners. When designed with full multimedia functions and combined with user control of the speed of presentation and the number of repetitions, on-demand instruction resembles the customized playlists and other individualized programming options that characterize how members of the digital generations consume media.

Multimedia Learning

When computer technology (hardware and software) advanced to the point where graphic user interfaces and network connectivity became standard components of personal computers (about the mid-1990's), they had also advanced to the point where they could display multimedia content (high-resolution graphics, and audio, video and animations) as well. From the second decade of the 21st century, it is hard to recall the pixilated monochromatic graphics and beeps that were multimedia in the early days of educational computing.

The multimedia capabilities of computers allow computer users to see (both static and moving) images, hear sounds, and control information in a ways that were not previously available. This capacity also makes the information and interaction more natural to humans than print-mediated information and interaction. Print-based information and interaction is limited to vision (Braille and text read aloud obvious exceptions) and is limited by the literacy skills of the writer and the reader and also by the shared language of the reader and the writer. By comparison, multimedia information and interaction allows multiple channels to be used in conveying information and multiple senses to be used in receiving information, it requires little special training to create and consume, and even individuals with different languages can use gestures to begin communicating via multimedia.

Learning through multimedia has focused much research of learning science; incorporating that research to designs of on-demand instructional will increase the performance expectancy associated with the use of the materials. First, learners do pay increased attention to information that is presented in multiple media compared to information that is presented only as text, but motivation and the level of one's cognitive engagement also affects attention. The "teachable moment" aspect of on-demand instruction is real and increases attention.

Second, multimedia can be understood as a progression of modes, each better than the preceding. Simple text is the least effective. Text supported with graphics is better. Text and graphics presented together is better. Visual text, graphics, and an audio version of the text is better still; and a natural voice reading the text is better than computer-generated voice (Mayer 2005).

Third, options that allow the user to control the presentation (pause, rewind, fast forward) as well as navigation aids and content cues are associated with improved attention to and memory of multimedia resources. This effect is observed when the cues and controls are well designed and do not introduce extraneous cognitive load or draw attention away from the content (Hede and Hede 2002).

In addition to worked examples, multimedia learning has proven effective in providing context or experience that can make other experiences more relevant (Mayer 2001). Providing students with an animation of complicated cellular processes in biology, for example, can provide a scaffold to facilitate understanding of textbook description of the processes and also facilitate understanding of observations in laboratory exercises. Such a presentation can be understood as providing students with tacit knowledge of the phenomenon under study. In this example, the multimedia experiences can also be interpreted as decreasing the extraneous cognitive load of the traditional instruction.

Open Resources

Open resources include all of the intellectual property in the public domain as well as that published under a license that allows users to use and copy it. It is a misconception that one can use open resources for any purposes; for example, certain Creative Commons licenses restrict users from including the property in commercial products. In general, the community of authors and programmers who produce open resources define them as resources which users can:

- *reuse* in a manner consistent with the terms of the copyright without obtaining explicit permission
- *revise* to meet their specific needs (in many cases, it is expected that these derived works also be published under an open license)
- *remix* with other resources
- *redistribute* the work in a manner consistent with the terms of the copyright without obtaining explicit permission

There are several coincident trends in the digital information landscapes that contribute to the open resource movement. First, there is a culture of "information wants to be free" among many leaders and workers in the information technology industry. The large and active open source community of programmers who produce and publish open source software demonstrates this culture. Second, the tools commonly labeled web 2.0 made web publishing easy and quick. Using these tools any person with access to the Internet can create content and disseminate it globally. Third, the growth of broadband Internet connections makes transferring large files, including multimedia files, reasonable. The sharing ethic that is common in academic communities combined with widely available technology for creating produces an environment in which open resources can be very effectively developed and deployed.

For generations, the textbook has been the dominant source of curriculum materials, and a large and lucrative industry publishes and markets those resources. As the open resource community expands and more experts are creating and publishing open resources, there is increasing interest in replacing the very expensive (to obtain and sustain) physical textbooks with open resources. In many cases, these resources increase performance expectancy as they can be easily and legally modified and distributed to meet the local needs.

In addition to making use of open resources, schools are increasingly creating their own open resource communities. As they create and share resources, they build a curriculum repository, which contains open resources that have been developed for the community. Educators with access to a curriculum repository are free to use any items that exist there, but the expectation is that individuals give as much as they take.

Managing Classrooms

Educators spend their days engaged in two distinctly different types of professional activities. Those involved with planning and creating learning environments and those involved with managing the learning environments. Both are information-rich activities, but the nature of the tasks is quite different.

When planning learning environments, educators must manage resources for teaching such as texts, multimedia files, and links to web sites as well as products of learning such as completed tests, papers, and physical products. We have already encountered the virtual classroom as essential technology infrastructure in the 21st century school (see pages 123-131). Virtual classrooms facilitate several tasks related to managing and interpreting information and supporting interaction to facilitate deeper, active, and authentic learning, and these contribute to increasing performance expectancy.

Designing learning environments is a wicked endeavor, so any solution can be evaluated and improved. Educators change their lessons and their tasks for a wide variety of reasons. New resources sources become available, others are no longer available; context changes so previous approaches become inappropriate or insensitive, formal study and serendipity lead educators to revise tasks and prompts and scaffolds that support learning. By using digital versions of resources and materials, teachers reduce the effort necessary to make changes that increase the performance expectancy of the materials.

Metadata is information about the information and it is encountered in many forms. In a word processing file, the author can add keywords to the document properties, and these become a searchable feature of the document. Adding metadata is frequently called "tagging" as it is an accurate summary of the process. By adding metadata to digital files containing curriculum materials, educators can further improve both the effort expectancy and the performance expectancy, as it is easier to find the materials that are best suited for a particular circumstance.

In addition to facilitating the management of information related to teaching and learning, technology can increase the performance expectancy related to management of other data. With technology; budgets, attendance records, grades, and other data can be managed with increased efficiency, which increases the performance expectancy of technology in educational work.

A common problem for educators at all levels is the student who does not complete assignments for any number of reasons. Once a student realizes work is missing (or more likely a parent or other adult realizes work is missing), the student arrives at the teacher's door or desk and asks, "What am I missing?" Teachers who use digital grade books and who announce projects on an online venue are well prepared for managing these situations with efficiency. Educators who seek a quick and easy way to immediately support their work are well advised to begin a homework blog to create a permanent record of assignments and expectations.

In the 21st century, social media has become a dominant venue for individuals and organizations to create and maintain an online presence. While Facebook is the largest of these and many schools maintain a presence on Facebook, other social media platforms have applications for education as well. Twitter is described as a microblogging tool as posts are limited to 140 characters. Many educators are reluctant to adopt Twitter as individuals' and groups' mistakes on posts are frequently the cause of embarrassment. When well used, however, Twitter can be very useful and effective; the following are examples of educators who have effectively communicated with communities:

- A math teacher who tweets pictures of solutions so students can follow her and see the images on their mobile devices or computers, and the feed is embedded in her virtual classroom.
- A middle school principal updates field trip progress so parents know their children safely arrived and they know estimated times of arrival back home.
- An athletic director of a high school who sits with officials during storm delays. Student athletes are not allowed on fields until 30 minutes after lightening is seen or thunder is heard. Whenever the "storm clock" restarts, he tweets an update, so spectators and coaches who are waiting in safety get updates.
- A guidance counselor who tweets when colleges are visiting the school, when financial aid forms are due, registration deadlines for SAT's, and similar items necessary for college admissions.

Although presented as independent factors, effort expectancy and performance expectancy are related and the level of one can affect the level of the other, especially in managing data. Consider, for example, a student information system (SIS), which is used to manage grades, attendance, and other details of students' enrollment. Using an SIS can increase performance expectancy coincident with improved efficiency of this task. If, however, the SIS is poorly designed or if the networked used to access it is unpredictable or unreliable, then effort expectancy will decrease contributing to a decrease in performance expectancy as well. As a result, technology-planning decisions must focus on ensuring that both effort expectancy and performance expectancy are increased. Further, decisions and initiatives must be evaluated though the lenses of effort and performance expectancy, this means the complaining teachers are permitted to complete the planning cycle (see pages 110-112) must be in terms of improving the ease of use or the usefulness of systems.

SOCIAL INFLUENCES

Educators are social beings (just as all humans are) and so the social environment in which they live and work is meaningful and influences their intentions and their behaviors. When school and technology leaders take steps to ensure that individuals perceive other individuals (especially those who are respected and perceived to be doing similar work) to be using technology, then there will be increased use of technology.

Social psychologists recognize three types of social influences. Social influences motivated by compliance are typical of settings in which individuals are obligated to act in a defined way to gain reward or avoid punishment. Educators who are required to use a particular online grading system may comply with the request, but not use the advanced features of the reporting system. When individuals feel a strong identity with another individual (or with a group of individuals) then the individual will seek to model the actions of that individual or group. The social influences resulting from those identifications tend to be stronger than the social influences of compliance. The strongest social influences arise when the individual internalizes the social influences, so they are perceived as natural and the individual holds the same expectation of others (Aronson 2003). The most sustained technology efforts in schools are those that follow faculty internalizing the social influences, although all of the types of social influences are commonly encountered within faculties.

Technology for Deeper Learning

Social influences are grounded in the vision of "screen time" that is shared in the school culture; it is subject to change as understandings of educative experiences change. It is also affected by the nature of the information and interaction that can be easily accessed through the network infrastructure. Systems that allow limited access to blogs, wikis, and other tools for technology-mediated interaction can severely limit an important aspect of human learning. School and technology leaders, must forge collaborative relationships to ensure that the technology supports appropriate teaching and learning.

Arthur Chickering and Stephen Ehrman (1996) concluded effective technology-rich learning cannot occur by either technologists or educators working in isolation. Technology must be selected and installed and used promote and facilitate seven principles of good practice. Technology can be must be installed to support these purposes:

- encourage contact between educators and students
- encourage reciprocity and cooperation among students
- facilitate active learning and performance as demonstration of new learning
- be a venue for prompt feedback
- allow students to spend time on meaningful cognitive tasks
- communicate high expectations
- respect diverse talents and ways of learning

In Chickering and Ehrman's list, the technology necessary for a school implicit, technology must be designed to provide a venue for interaction as well as information and the tasks must be engaging. Technicians and teachers must share the expectation that technology support deeper learning.

Rapidly Evolving Expectations

The social influences arise from the collective understanding of what one should do. A former teacher and then colleague who had a long teaching career (sufficiently long that one of his students was the grandchild of another of his students) was fond of saying, "Education is like the weather in New England... it changes every 10 minutes." He was commenting on the seemingly random direction of educational initiatives that he had seen advocated over his career. In chapter seven, this phenomenon is referred to as horizontal reform. Given this observation, the changing (and recurring) expectations arising from educational reform and school improvement efforts are a familiar characteristic of education.

In recent decades, many governmental agencies, non-profit organizations, and professional organizations (for educators and others) have produced a variety of curriculum standards. Ostensibly, these are produced for the purpose of providing an objective definition of what constitutes appropriate topics for students throughout their school careers. These become the default expectations for what teachers will include in their courses. Most recently, the *Common Core State Standards* (CCSS) is the collection of standards that have been criticisms leveled against those standards (see page 4), which is leading some to predict that they are to be soon-replaced.

The evolution of technology is known to follow an s-curve (see pages 28-29). When a technology reaches the end of it's s-curve, it tends to be replaced by a new technology. The evolution of computers and digital electronic technologies has been marked by straighter s-curves than previous technologies. This suggests these technologies change more rapidly than previous technologies. A similar pattern can be observed in patterns of technology adoption; straighter s-curves demonstrate new platforms become widely used in months compared to years or even decades for early electronic media

(Hanneymr 2003). This fact is complicating the social influences of technology acceptance.

Whereas 20th century educators were able to develop their baseline professional knowledge as undergraduate students and then complete and update all aspects of their professional knowledge through occasional review over years, the 21st century educator will be continuously refreshing his or her practice with very brief periodicity. In the journal that I kept when I was a student teacher in 1988, I recorded the advice of one of the teachers in whose class I taught, "Every few years, whenever they buy new textbooks, you have to go back through your plans and update them." In 2000, I recorded the observations of a teacher who was enrolled in a summer institute to design a system of electronic portfolios. In the first year of the institute, she had learned to create portfolios to be put on compact disks, and at the second institute she was developing the process for putting them on the World Wide Web using hypertext markup language. She observed, "So now, every year or so, we need to update how we do things for new technology." When I was observed as part of our teacher evaluation program in 2011, my principal noted that the lesson I had presented was not the one I had indicated in the previous week the students would be studying. My response seems to capture the extent to which educators must review their curriculum and instruction. I explained, "If I don't—at least once a week or so—learn about something one day and share it with my students the next day, then I am not doing my job. You happened to be in my classroom the afternoon after I learned something good when eating lunch."

As school and technology leaders look to design professional learning to support educators who are active in the on-going work of reinventing technology-rich classrooms which will be permanent part of the 21st century education paradigm, it will be necessary they incorporate ideas such as Deuze's (2006) participation, mediation, and bricolage and Furr, Ragsdale and Horton's (2005) formative and naturalistic evaluation, which—compared to outcome-based evaluation—recognizes more relevant factors and allows for more diverse sources of performance in teacher evaluation than test scores. Further the professional development will be characterized by non-linear experiences for teachers and by preparing teachers to engage in nonlinear planning for the teaching and learning in their classrooms.

The professional development must provide educators with the opportunity to develop awareness of the emerging technology skills needed by and for their students as well as the origins of those skills from within the changing sociocultural context. Emerging trends that are redefining the skills and knowledge necessary for economic, political, and social success for the citizen of the 21st century have been reviewed; such changes will focus the attention of educators on the context of the changes necessary in 21st century classrooms; understanding such trends will help educators know why new curriculum and instruction is important. Just as classroom will become places where deep, active, and authentic learning is common, those qualities will mark professional learning for educators as well.

Collaborative Curriculum

The work of creating and implementing authentic curriculum is far more complex and unpredictable than creating and implementing curriculum that comes from textbooks or other highly prescriptive sources. As has been explained in previous chapters, authentic curriculum is designed locally but is grounded in discoveries from cognitive and learning science. The local curriculum is specific for the populations, goals, and resources of the community. These complicating factors make the task of creating and updating curriculum one that is well suited to collaboration. Commenting on the need to organizations become more flexible in planning, Douglas Thomas and John Seeley Brown observed:

> The new culture of learning is based on three principles: (1) The old ways of learning are unable to keep up with our rapidly changing world. (2) New media for are making peer-to-peer learning easier and more natural. (3) Peer-to-peer learning is amplified by emerging technologies that shape the collective nature of participations with those new media (2011, 50).

Building on the culture that has facilitated the growth of open resources, leveraging the technologies that support that community, and reflecting the new learning culture, many school and technology leaders increase social influence and increase performance and effort expectancy by organizing curriculum repositories. The online curriculum repository becomes an invaluable resource as the focus of a professional learning community in schools committed to technology-rich deep, active, and authentic learning. The repository is an online space where teachers go to find and share links to appropriate online videos (and other media), locally created worked examples, templates for planning and assessment, ideas for prompts, and similar resources to support teaching and learning.

Many educators find the work of managing curriculum in the 21st century has changed from creating to vetting and editing. In traditional classrooms, teachers take much time preparing materials for the classroom: notes are prepared, papers are copied, books are collected, videos are cued, and materials unpacked and set out for students to use. In addition, teachers prepare to assess students' learning by selecting and transcribing questions to compile tests, and copy and disseminate the tests, and then preparing grades to be reported to parents and others.

In 21st century classrooms, much of that work is digitized, and so can be completed with less effort, and many of the resources are digital so require less preparation. A teacher preparing a unit of study will likely make changes to resources that already exist and will investigate resources to ensure they align with curricular goals and are age-appropriate. Once appropriate resources are identified, many educators turn to the task of creating outlines or other scaffolds to ensure the resources fall within students' zone of proximal development. These resources, both the curriculum materials and the scaffolds, comprise the growing curriculum repository available to a faculty.

In today's vast information landscape, the work of building a curriculum repository can include vetting resources as much as creating resources. Fortunately, vetting resources is a task that is very easy to share, and when local communities control the curriculum repository and the users have a relationship with the contributors, they tend to have greater confidence in the resources. Many communities will create a course on the LMS that serves as a curriculum repository. Teachers make contributions and revisions to help build the collection and tags and other metadata to help organize and manage the collection.

Experience to Model

As educators collaborate to build and manage the curriculum repository they will be creating digital information, as well as interpreting and analyzing information. As they do, they will be gaining experience using information technology tools to build an online community of practice, thus experiencing the type of technology-mediated interaction they will create in their classrooms.

Bonnie Nardi (2005), a scholar from the University of California, Irvine, observed that in settings in which communication occurs via technology (specifically, Nardi studied instant messaging), three factors are associated with sustained interaction. In settings in which an individual feels a member of the group (affinity), a desire to contribute to the activity of the group (commitment), and is motivated to know and understand the activity of the group (attention), an individual is more likely to sustain communication within the group. Sustained communication is necessary for the social construction of knowledge and for creative and complex problem solving within a group, therefore communities must encourage "the creation and renewal of social bonds of affinity, the establishment of commitment, and the capture of attention" (Nardi 2005, 125). The experience of contributing to a curriculum repository prepares educators to be active models of technology-mediated professional interaction for their students.

Charlotte Gundawardena, Constance Lowe, and Terry Anderson (1997) observed that participants in online discussions engaged in social interaction for five purposes: a) sharing information, b) discovering dissonance, c) constructing new knowledge, d) testing new knowledge, and e) applying new knowledge. In their data, sharing knowledge accounted for the greatest number of interactions (almost 200 postings compared to fewer than 10 postings in the other purposes). This suggests that social interaction in online environments, even when focused on content, does not necessarily lead to deeper understanding. Through prompts that facilitate diverse interaction, teachers can encourage more interesting and engaging online interaction.

The conclusions of Nardi and Gundawardena, Lowe, and Anderson suggest that a curriculum repository must extend beyond simply pointing colleagues (and thus students) to sources of information. The prompts used by teachers to encourage and organize interaction in virtual spaces influence the nature of the experience for students. These should be the focus of discourse in the curriculum repository. Teachers should share prompts, their observations of the effectiveness of the prompts, and engage in discussion for the purpose of improving the prompts.

Through their own active learning about the transition from in-person to online social dynamics, educators can facilitate the transition for themselves and for their students to create environments to which students feel affinity with and commitment to a class as a social group of learners. Once protocols to guide these social interactions are established, educators must participate in the interactions and model the kind of interaction they expect, thus becoming mentors for apprentices.

PROFESSIONAL DEVELOPMENT

Recognizing that teachers are flexible professionals who specialize in using technology to support teaching and learning require on-going opportunities for professional learning is essential for school and technology leaders. This professional learning will be characterized by a mix of self-selected and self-defined learning (based on one's expertise and understanding of current need) and new ideas, which extends and expands teachers' understanding of their work. Several models for professional development are appropriate. An effective professional learning program will include each.

Furr, Ragsdale, and Horton (2005) suggested that teacher evaluation in the increasingly complex world should be formative and naturalistic. They claim many teacher evaluation programs focus on the outcomes of technology-rich teaching and learning and thus avoid the complexities of authentic learning that reflects the dominant culture and ignore essential aspects of human learning, society, and the role of information technology in that milieu. In arguing leaders evaluate teachers in a naturalistic manner for formative purposes, they suggested it will account for factors related to the unique experiences of the populations in the classroom and that it will be designed to improve the curriculum and instruction for those populations. They argue for professional development that prepares educators for this setting will prepare them to "assess the expected and unexpected consequences [of technology] and then to adjust their teaching to employ technology as a positive, cognitive tool" (285). Professional development to meet this vision will be organized in several ways.

Awareness Presentations

Because information technology (hardware and software as well as network resources) changes so quickly and new tools are developed and refined so quickly, it is likely that options exist that even connected educators are not aware exist. The purpose of an awareness presentation is simply to introduce a technology or strategy to an educator (or group of educators). After an awareness presentation, those educators who are interested in the topic will have access to resources for further exploring the topic. Awareness presentations are characteristically:

- *very brief* (10 minutes or so)- A "show-and-tell" session included in the agenda of a faculty meeting is common, as is a round robin with several awareness presentations set up in one place and groups rotating through the individual presentations.
- *allow for reflection*- Because the new tools introduced in awareness presentations may be perceived as unnatural to some teachers, the chance to hear others' perceptions is a chance to connect with the new ideas and tools. Of course, the opposite may occur as well, and reflection may be the chance for educators to criticize new ideas or tools. Focusing discussion in awareness presentations around questions of effort expectancy (e.g. What will be easier if we use this?) or performance expectancy ("What will this help us do better?") tend to prevent negative reactions in awareness presentation reflection.
- *include options for further exploration*- By design, an awareness presentation gives the audience minimal (even no) experience actually using the tool. Because of that, there must be a way for interested audience members can learn more about the tools after the presentation ends.

Tutorials

Tutors are individuals who provide instruction to learners, and this type of support is helpful to those learning to use new technologies. Many are familiar with math tutors who review with students how to do certain problems and who observe and troubleshoot the student's work. Tutorials for technology tools are similar. Whereas traditional tutors meet in-person (and they can be effective to teach how to use technology tools), many technology tutorials are given via recorded presentation and the audience has control over when it is viewed and how often it is viewed. Tutorials are generally:

- *intended for individual or small-group support*- Presenting a tutorial to a large group is seldom effective as the tutor goes too fast for some while others are far ahead of the group.
- *Best when provided just in time* as the learner needs the information. This is one of the advantages of using information technology to deliver tutorials. Users are relieved of the responsibility of remembering (or otherwise expending cognition) and the users tend to be motivated so they pay greater attention and retain more of the information.

- *Best for step-by-step tasks* in which the steps are either correctly performed or incorrectly performed.

Tutorials are in many ways like the on-demand learning that is known to improve the performance expectancy associated with technology acceptance. The same principles regarding multimedia design that improve the effectiveness of those materials for students can be applied to supporting educators' learning as well.

Institutes

Institutes bring a group of educators together for an extended time (typically measured in days) so they can participate in an intensive and immersive experience. For several days, educators dedicate extended time to (with guidance) conceptualize, draft, refine, and prepare classroom materials. This is very useful when first adopting an instructional model, and time is needed to prepare a foundation of materials to support initial efforts in the classroom. When initially adopting virtual classrooms or when creating the initial curriculum repository, school and technology leaders who organize an institute are likely to positively affect all of the factors associated with technology acceptance.

Institutes combine several features:

- *consideration of theory*- Especially in those institutes that include graduate credit for participants, the curriculum of the institute includes reading and experiences that help the participants understand why they are learning new models.
- *expectation of product*- Participants should leave institutes with resources that will help them as they return to the classroom. Lesson plans, curriculum materials, and handbooks are all examples of the type of products they take with them when the institute ends.
- *a cadre of leaders*, ideally comprising those dedicated to the theory and experienced in the classroom to guide participants as they explore the theory and prepare for the practice of the model.
- *on-going support* to help participants assess the work and reflect on necessary changes after they return to the classroom.

Support at a Distance

Several technology tools can be leveraged to extended and expanded the potential learning communities for educators. Educators can collaborate with the same virtual classrooms for their own professional learning that are used for their students. Many observe that this is perhaps more important than in-person professional development, as teachers who are interacting at a distance for their own professional learning are experiencing the potential and the difficulties of using virtual classrooms as a learner.

School and technology leaders have a range of technology tools to support synchronous professional learning. Of course, these tools are not unique for professional development for educators; they have applications in teaching and learning with students, and are widely used for training purposes by software and hardware vendors and other organizations working in educational technology fields.

- *Chat*- A chat room is a web page that contains two text boxes, one seen by the individual user and another shared by every user who is logged on to the chat room at a particular moment. When one enters text in the box on his or her screen then clicks the "send" button, the message is instantly displayed so everyone in the chat room can see it. Chat is built into most LMS, and chat is frequently used to interact with technical support providers. When designing curriculum, chat is often useful for brainstorming as participants can contribute with little attention to editing. The transcript can then be reviewed for ideas worth developing.
- *Screen sharing*- As the title suggests, screen sharing is a technology that allows one to share their screen with others. Typically, the presenter will initiate the sharing and send a link to the audience so that they can point their web browser to a web address that displays the presenter's screen. This allows the presenter to demonstrate software or web site, or for a dispersed audience to focus attention on a single display. This can be combined with chat or audio over the Internet or the telephone for even more interaction. Screen sharing is often used to demonstrate new information platforms.
- *Video conferencing* in which participants see and hear each other. Some platforms also allow for screen sharing and some allow for different users to take control of the mouse and keyboard of the presentation computer remotely. The best way to connect to a video conference is to be sure your computer (or mobile device) is connected to a wifi network (the systems do not work well over cellular connections).
- *Social media*- Facebook, LinkedIn, and Twitter (among other platforms) are all spaces where groups of educators have formed extended communities of practice. Participants share resources and ideas and provide on-going commentary on trends, tools, and events.

All of these tools have build incorporated into webinars, which are widely used in business. K-12 education has been relatively slow to adopt webinars as the technology for providing high-resolution images of screens, high-quality audio, and chat functions to groups the size demanded by educational purposes tends to be very expensive. In addition, the systems tend to be difficult to use because of the many functions available in the interface.

Chapter 7
Developing System Capacity

If we accept that education is a technology that addresses the wicked problem of preparing young people to participate in a quickly changing global social life, then it is reasonable to conclude that no single action (or set of actions) will suffice to permanently prepare a school for the work. Teachers, faculties, schools, and the entire education system will be prepared for digital generations only when educators accept the challenge of reinventing (and continuing to reinvent) their practice. The reinvention also requires that political leaders recognize the nature of learning. Educational paradigms that focus on instruction and measure achievement with standardized tests are naïve and ignore the facts of human nature and the skills necessary for full participation in the emerging culture.

Especially in this century, education has become the focus of much political attention. Government agencies, politicians, and philanthropists are all much more influential in determining educational policy and practice than they were in previous generations. Neal McClusky, a policy analyst for the Cato Institute's Center for Educational Freedom observed the effects of *No Child Left Behind* (NCLB), a law which increased the role of the federal government in education, and concluded, "the signs were clear that NCLB was just another educational reform initiative rich in rhetoric, but fundamentally bankrupt." He used this as an example of the dangers of politically-driven educational initiatives, "by its very nature, a system of government schools for which everyone must pay but which only the most politically powerful can control is doomed to failure" (2007, 95). The politically powerful appear to be those who are the least aware of the nature of human learning.

Following the evidence presented in the previous chapters, it is reasonable to conclude that school planning, especially that planning related to teaching and learning, will follow nonlinear but directional paths. This work will be grounded in cognitive and learning sciences but recognize the local circumstances that affect how populations are served and resources are used. Such systems will be designed to allow access to technology and decision-making and agency for all participants. Educators who actively engage in the process or designing such systems are likely to create the future of their institutions rather than to have it imposed by outsiders.

Throughout this book, education has been presented as a human endeavor that is undergoing a radical change from what it has been for many generations. As we have seen, the experience of education for students is becoming more active and authentic; the work for educators is becoming more collaborative and curative and innovative. In this final chapter, attention turns to changes in school and other organizations that will support teachers as they use technology to create deeper, active, authentic teaching and learning. A recurring theme in this book has been adaptability; we seek to design classrooms that nurture learning that can be adapted to unforeseen circumstances. We seek to support educators as they adapt to changing technology and curriculum needs.

Allowing Vertical Reform

Educational reform tends to follow a cycle that is familiar to many: First, an initiative (supported with little or dubious evidence from the learning sciences) is introduced and implemented (with little or dubious support and rationale). Second, problems with the initiative appear. These can originate from poor or incomplete implementation or support, discrepancies between the practices and human nature, or other difficulties. Third, the initiative is recognized as failing, but remains in place (or is replaced with previously used methods under the vocabulary of the initiative). Fourth, a new initiative replaced the old and the cycle repeats. Frequently the choice of next initiative and the time devoted to any initiative depends on the availability of grants to support the work. Following this model of reform, educators can appear to be working to improve curriculum and instruction while avoiding implementing any new practices. This also allows educators to abandon any initiatives that force them to resolve any challenges to their existing practice.

The result is what can be called horizontal reform. Schools are perpetually beginning new practices, and none is ever allowed to have deep influences on pedagogy and student experiences. A colleague who is a known cynic commented on my version of horizontal reform and observed, "It is probably best that none of these horizontal reforms ever gain traction. I have never seen any that is as effective as they claim."

Barbara Cambridge (2008), a scholar who has held many leadership positions in professional organizations, has defined vertical reform as that which is sustained and allows educators to internalize shared understanding of effective teaching and learning. When this occurs, there are strong social influences. In describing how school leaders can facilitate these changes, Cambridge suggested the process includes redefining purposes and strategies, including socially dynamic and active planning and designing activities that reflect deep, active, and authentic learning. In addition, Cambridge suggested building upon existing successes, so that (for example) existing efforts to establish improve deeper, active, and authentic teaching and learning in a particular school becomes the foundation for increased and expanded authentic learning. Systematic change requires educators adopt the courage to change and possess both patience and impatience. Patience because all change takes time to become established, and impatience in the face of excuses for not engaging with the reform.

Toru Iiyoshi and Cheryl Richardson (2008), both from the Knowledge Media Lab of the Carnegie Foundation for the Advancement of Teaching, noted that educational transformation, especially in the digital age, is characterized by "the impulse is to keep moving to newer and niftier 'gadgets' without learning thoroughly from what worked and what did not" (338). Reflection, it seems, is no longer among educators' tools as related to their own planning. Iiyoshi and Richardson conclude, "In many cases, familiar educational tools and resources would be more effective if educators, in partnership with students, would simply try to devise ways of using them to deepen students' understanding" (338). As educators adopt the planning that will support their understanding of familiar (and new) tools and resources, they will abandon the horizontal reform, and begin to reform vertically, in that change in capacity and culture and reform activity based on evidence and reason becomes systemic. Vertical reform is opposed to horizontal reform where initiatives are replaced before being fully implemented.

Updating Assumptions

As has been established, there are several assumptions about teaching and learning that formed the foundation of the 20th century educational paradigm. Sawyer (2006) suggested these assumptions included the idea that what should be taught (the curriculum) is well-known, instruction is the best method for delivering the curriculum, educators understand instruction, curriculum should remove complicating factors, and testing as the best method for measuring learning. Sawyer also recognized that incompatibilities between those assumptions and emerging discoveries from the learning sciences would cause difficulties for educators. Evidence of the effects of the difficulties is seen the failure of horizontal reforms. Vertical reforms are grounded in the cognitive and learning sciences.

Many of the educational reform efforts in recent decades have only focused on changing curriculum and instruction and the underlying assumptions have been left unchallenged by reformers. Because those efforts have been informed by what we now know to be inaccurate knowledge of teaching, learning, and schooling; the ambiguous and inconclusive results of reform efforts are not unexpected. Educational communities can expect steps taken to improve education that do not compel the community to reevaluate the assumptions at the foundation of curriculum and instruction in light of new discoveries will be found unsatisfactory.

Vertical reform is a paradigm changing endeavor. While many educational initiatives are promoted as paradigm changing, few are. As used today, the concept of a paradigm can be traced to T. S. Kuhn, a philosopher of science, who defined four components of a paradigm; unless all of these are changed, the paradigm is not changed.

Kuhn (1970) called the first component of a paradigm the *symbolic generalizations* and these include the natural laws that are the foundation of the science. Kuhn's examples from physics included Ohm's Law, which describes the flow of electricity through circuits. Because education is a social endeavor as opposed to a natural science, it is more difficult to recognize the symbolic generalizations in education than in physics, however. Education has been built on the assumption that educators know how to transmit curriculum into students' brains, and embedded in this are symbolic generalizations about how brains work to construct and apply knowledge. The learning sciences and cognitive sciences are elucidating the natural laws and symbolic generalizations relevant to this component of the Kuhnian paradigm for education, and these are much different than the symbolic generalizations that informed previous generations of educators.

The second component of a Kuhnian paradigm is the *metaphysical paradigms,* which "supply the group with preferred or permissible analogies and metaphors" (Kuhn 1970, 185). In the 20th century, the idea that the world is filled with well-defined information and skills representing necessary human knowledge that can be transferred into, stored within, and applied by human minds is a metaphysical paradigm that has been applied to education. This mind-as-a-container metaphysical paradigm is being challenged by evidence that human knowledge is a social construction. Another metaphysical paradigm that has influenced K-12 schools is that technology is a neutral aspect of information and of culture, so we can predict the outcomes of human interaction with technology. This also is challenged by recent scholarship.

Values are the third component of paradigms and include the broadest ideas that connect the group that study and practice in the field. In science, values include valid and reliable data and open sharing of data. In education, an example of a value is the assumption that the purpose of schools is to ensure students become educated so they can fully participate in the economic, political, and cultural life of the society. Another is that adults understand and have a role in preparing young people to participate in society. This appears to be the one aspect of the educational paradigm that is least challenged in the transition we are observing in education, but some values in education (such as the role of authority) are being challenged.

The final component of a paradigm includes the *exemplary practices* that individuals who are entering the field are expected to master. In science, Kuhn suggested, these are identified in textbooks through the problems students are expected to solve. As part of educators' professional preparation (and his or her on-going professional development), they are introduced to and expected to gain experience using various teaching and evaluation methods. A sign of the need for a paradigm shift is the series disparate and mutually exclusive practices that have been promoted to educators; political and social leaders perceive 20th century education as insufficient to meet the needs of 21st century society, but they are unsure of the necessary curriculum and instruction, so they advocate for practice after practice in search of the elusive exemplary practice. Once the paradigm shift in education is complete, it will become clear what learners should experience and how teachers should design classrooms.

Critical Consciousness

Paulo Freire, an educator who worked in Brazil in the 1960's, is well known for several essays including "Education as the Practice of Freedom" and "Extension and Communication" (Freire 1974). In these works, Freire argues that meaningful learning occurs when the learner reaches critical consciousness which enables the learner to reflect on and understand not only what they know, but why they known it and how they know it. For Freire, education is a problem of negotiating information which is created by humans and within a social context; purposes, frames of references and reasons all contribute to dialogue between human and the information they construct together. Referring to the on-going work of answering questions that arise during instruction, Freire observed, "It means making a new effort, in new situations, in which new aspects which were not clear before are clearly presented to the [learner]" (1974, 151).

In the recent history of educational reform, very many proposals have been presented as vast improvements over current practice. Rarely are these presented in a comprehensive manner including the purposes, frames, and reasons that Freire argued is the basis of education. Once educators (both individually and collectively) have articulated their understanding of education and reconciled that with the facts from learning science, they are prepared to accept or reject any of the myriad of educational reform proposals.

The future of education planning necessitates each educator (both individually and as a community member) develop critical consciousness that empowers each individual and group to respond with reason and evidence to proposed pedagogy. As assumptions are updated and practice is reinvented, educators will apply their critical consciousness to update their understanding of all four components of Kuhn's paradigm.

TECHNOLOGY PERSONNEL

Technologies, including educational institutions, are modular and undergo structural deepening. With the arrival of computers and computer networks, the technology support module was added to schools. This module includes personnel who support both the technology infrastructure as well as the use of technologies in learning and management. Early in the history of computers in schools, many individuals possessed sufficient skill to both support the relatively simple infrastructure and support the use of computers in schooling. As the technology became more sophisticated and as the pedagogy became more sophisticated, it became necessary to have different individuals to support the infrastructure and the use.

Current Roles

In schools today, most professionals who work with technology typically fall into one of six groups. The descriptions of the duties assigned to these individuals illustrate the range of tasks necessary to manage and use the very complex information technology infrastructure encountered in a typical school and the scale of the system that requires support:

- *technicians* who perform basic installation and maintenance on devices, this includes providing first level help and frequently interacting with technical support services from vendors and manufacturers
- *network administrators* who install and manage servers and other network infrastructure, these individuals frequently have certifications from professional organizations and their expert knowledge is often provided through contracted consultants in schools
- *technology coordinators* who support administrators making decisions related to computers and information technology in the school, frequently individuals in this role will translate technology needs and decisions for educators and educational needs and decisions to technicians and network administrators
- *teachers* of programming, multimedia, and productivity applications classes, these individuals are licensed teachers
- *technology-savvy teachers* who teach subjects other than computers, but who use technology as an instructional tool
- *technology integration specialists* who work to support teachers to use technology in their classrooms, these individuals are usually licensed educators as well who have transitioned to faculty support and development

In many schools, of course, the boundaries between these jobs blur and a single individual may serve in multiple roles either formally or informally. Just as it is not appropriate to have a teacher providing network support, it is not appropriate to have a network administrator making educational decisions.

Technology Stewards

Unless technology has become embedded into the culture of the organizations, the role of the technology steward is not being fulfilled; technology stewards have only recently been defined as a unique technology leader (Wegner, White, and Smith 2009). Technology stewards are defined within the communities of practice. Communities of practice (CoP) is a concept developed by Etienne Wenger and Nancy White and their collaborators; the idea has influenced organizational researchers and planners since the turn of the century (Wenger 1999). Each CoP is defined by a group of practitioners who share a common endeavor and who also share a collection of practices to accomplish the endeavor. CoP's emerge when a group of individuals gather to share expertise to solve their common problems. While not originally intended as a model to promote interaction in technology-related fields, several strong connections between CoP and information technology have emerged.

The individuals who fill the role of technology steward often emerge informally. Technology stewards are neither the technologists who are responsible for technology infrastructure, nor are they exclusively the practitioners who use infrastructure provided by the technologists to accomplish the goals of the CoP. Technology stewards are leaders who discover, invent, and share the practices through which the technology can be used to accomplish the logistic and strategic goals of the CoP. Typically these leaders come from within the membership, and these individuals are deeply familiar with goals and work necessary for the community, and they seek technology that meets those goals and supports that work. Wenger, White and Smith observed, "Tech stewards attend both to what happens spontaneously and what can happen purposefully, by plan and by cultivation of insights into what actually works" (2009, 24). Further, technology stewards are empowered with more than just cursory decision-making authority; they understand the social systems of the CoP and the information technology that supports action within that social system and decision makers trust their assessments.

The key aspect of technology stewardship in 21st century schools is the emergence of the technology steward from the population of teachers. Cuban (1986) observed that administrators' directives to teachers to use radio, movies, and television were not followed because those leaders did not recognize and address important barriers to implementing the technologies in classrooms. Jeyaraj and Sabherwal (2008) argued that initiatives designed to increase the use of information technology in a diverse range of organizations were more successful if the effort originated with the users and their efforts to find tools to facilitate their work. When management mandates technology use, Jeyaraj and Sabherwal concluded adoption is inhibited as users adopt a passive stance and they use it only as mandated and not for other useful purposes. Because of the technology steward's knowledge of and experience within the CoP, members trust their assessments and recommendations.

Scholars and other observers recognize that technology adoption within an organization typically occurs in a linear manner, proceeding from the moment decision-makers first become aware a relevant technology exists and continuing through implementation until the technology is embedded in the day-to-day functioning of the organizations. Technology stewards play a role in streams of activity that can be interpreted as supporting perceived ease of use (or effort expectancy), perceive usefulness (or performance expectancy), and widespread adoption (or social expectancy).

Ensuring Ease of Use

The selection of technology can be a wicked problem for a community. The devices or software available are likely to have some features that are needed by the CoP but are underdeveloped in the available tools, while other features that are less important to the CoP will be highly developed in the tool. As a result, a poor technology selection can leave some important functions unmet, while introducing extraneous cognitive load for unimportant functions. By identifying which features are essential to both the strategic goals of the organization and logistic needs of the members, technology stewards reduce the potential for technology-associated noise.

The perspective of the technology steward from within the organization is vital insight in decisions related to ease of use for selection decisions. An educational faculty that becomes a CoP will comprise individuals who are reluctant to adopt technologies and individuals who are quick to adopt technologies. The relative size and influence of each of these sub-populations is considered when technology stewards advise selection decisions. In addition, the technology steward is often aware of the influences of other local or wide-scale initiatives or trends that influence that successful adoption of selected technology.

Once the devices are selected, the technology steward plays an important role in assisting with the installation. In many cases, that assistance is quite involved as deep understanding of the manner in which the installation was done is needed for the technology steward to both provide support for users and to provide feedback and information when technicians are troubleshooting malfunctioning systems.

The installation process of large technology systems includes testing of both planned systems and testing of fully installed systems; the technology steward is actively involved in these tests. Just as teachers "complain" about technology systems as part of the planning cycle (see pp. 110–113) and technicians must modify systems to accommodate the complaints, the recommendations of technology stewards during testing phases of installation must be given top priority. Consider the installation of a new fleet of laptop computers to be shared by teachers working in a school. The installation plan will include software, including web browser extensions, printers, and other devices as well as the addition of bookmarks and similar tools and shortcuts for both teacher and student accounts. Early in the installation, a single machine will the configured according to the plan, and the technology steward will test this device in a manner that teachers and students are known to and expected to use it. Any misconfigurations or problems that are identified on that test device will be addressed and resolved. Once the test device functions as expected, then that configuration will be replicated on the remaining devices. When teachers and students first use the devices, the technology steward completes the second phase of testing, and identifies problems that did not appear during single-device tests. Problems identified by the technology steward during testing of the full systems are given top priority for resolution as well.

Ensuring Usefulness

Technology stewards must be completely familiar with the purposes and the operations of the organization; they know what the members want to do and how they do it. It is within the context of what the organization does and how it does it that a technology steward perceives and interprets the ease of use and usefulness of technology. In terms used by hermeneutic researchers, the technology steward will interpret the usefulness and usability of any technology in light of his or her understanding of the community, and the technology steward will interpret the community in terms of the technology that can be leveraged for its purposes and practices.

It has been established that in the 21st century technology evolves rapidly as the industry pushes new devices into the market and as the market pulls new devices from the industry. Technology stewards pay attention to the evolving collection of devices on the market with an eye to discover those that meet the strategic and logistic goals of the community, and they share those discoveries with the community.

Everyday Use

The final stream of activity of technology stewards is to facilitate the work of embedding the technology-mediated practices into the normal operations of the CoP. When this process is facilitated by technology stewards rather than imposed by leaders or inefficiently discovered by members, Wenger, White, and Smith (2009) suggested the technology becomes part of the culture of the organization and the application of technology practices is more sustained. Technology becomes transparent to the members, and will be perceived as a natural and expected part of the operation of the organization in the way that adolescents' develop a sense of natural technology.

In educational settings, technology stewards often support the transition to everyday use by providing very direct and specific guidance and support for teachers. It is not uncommon to find technology stewards team-teaching with classroom teachers or even presenting technology-based lessons while the teacher observes. It is not uncommon to find technology stewards logging on to virtual classrooms to edit content with the teacher. It is also not uncommon to find the technology steward troubleshooting technology to ensure systems are easy to use and they are perceived to by useful by the members. In this troubleshooting, the technology steward represents the members to the technicians.

LEADERSHIP

Even in communities in which there is a strong technology steward, school administrators must play an active role in making and justifying technology decisions to all stakeholders. While making and explaining decisions is not an unusual role for school leaders, technology does appear to be a complicating factor that causes school leaders to be less comfortable in this role than when justifying decision in other aspects of school management.

Identify Trump Role

In chapter five "Technology-Rich Learning Environments," the technology planning cycle (see pages 110–3) was introduced. This cycle captures the different roles of technicians and educators in ensuring that information technology systems in schools serve the needs of students and teachers. The cycle specifies roles for both technicians and teachers.

Frequently, the goals of the technology leaders and technicians who are responsible for installing and managing networks are opposite to the goals of the educators. At different points in the planning cycle, different individuals have the more important role, and the decisions of the individuals filling that role must be given higher priority, thus they trump the others. School leaders must understand the cycle as well as be able to identify where in the cycle any technology is at a given point. School administrators find themselves in the position of directing technology leaders to follow the recommendations of teachers to redesign or directing teachers to follow the recommendations of technicians for using the technology. These decisions can only be made if the administrator clearly understands (or comes to understand) what has come previously.

Progressive Discourse

As a wicked problem, education—and thus the planning for education—appears different depending on many personal and societal factors that influence one's perspective. Stakeholder groups are commonly identified to categorize those with different perspectives on education; so we conclude teachers' perspective is different from parents' perspective which is different from administrators' perspective. Even within a stakeholder group, individuals are likely to have much different perspectives on educational problems (and solutions).

In situations where individuals with disparate perspectives are making decisions related to school management, the process can be highly politicized. Political problems are resolved by debate and discussion and—frequently—compromise. Political discussion and debate can be driven entirely by opinion; no participant is compelled to argue from accurate assumptions, arguments following dubious logic are permitted, as is any interpretation of evidence.

While decisions related to teaching and learning are subject to political processes, teaching and learning is fundamentally different from other aspects of school management. Teaching and learning is grounded in the natural phenomena of human learning, so participants resolving teaching and learning problems must hold accurate and true (or as nearly so as learning sciences can provide) beliefs about those phenomena. In this situation, discourse is more appropriate than debate or discussion.

Carl Bereiter (2002), the same educational psychologist who defined knowledge building, described the role of discourse in his profession. Through discourse, Bereiter claims, psychologists can "converse, criticize one another's ideas, suggest questions for further research, and—not least—argue constructively about their differences" (86). Discourse, then, must be organized around observable knowledge and dedicated to extending knowledge through research. For Bereiter, successful planning in education depends on discourse, but with variation that he refers to as progressive discourse, which is designed to better understand learning and to improve the manner in which teaching supports learning.

Progressive discourse begins with a well-established definition of specific actions that are described by the words being used. The nature of the actions captured in a term becomes a conceptual artifact. A well-defined conceptual artifact prevents stakeholders from applying too broad definition to the concept.

In political discussions, it is often desired that participants agree. In order to reach compromise, participants in political debate can diverge from common understanding; while agreeing on the language, the participants disagree on the action being labeled. (For example, touch typing lessons may be labeled "technology integration" to satisfy the expectation that technology be integrated or "making a poster" may substitute for composing essays to satisfy the expectation of projects.) While this step can lead to agreement, it does violate the focus of progressive discourse on improving conceptual artifacts. Clearly defined conceptual artifacts are the foundation of progressive discourse, so modifying a conceptual artifact to achieve political agreement is contrary to progressive discourse. Maintaining the integrity of the conceptual artifacts and the common understanding upon which those are built is more important in progressive discourse than political agreement.

All conceptual artifacts are incomplete and can be improved; the purpose of progressive discourse is to improve those concepts. In educational settings, improvement of conceptual artifacts is recognized at two levels. First, progressive discourse can lead to deeper and more complete and accurate understanding of the conceptual artifact. Educators learn to nuances of technology integration and project-based learning, for example.

Second, the manner in which the conceptual artifacts are instantiated can be improved. In the examples we have used, the practices technology integration and project-based learning as instantiated in classrooms and experienced by students can be improved. Both the theory and the actions that form the praxis of education can be refined and modified based on new learning, differing context, and different students.

All aspects of the conceptual artifact (e.g. how it is defined and measured, the conditions that lead to improvements, and who judges the improvements) are all subject to improvement through progressive discourse. Especially when progressive discourse is focused on wicked problems, the focus on improvement can extend to various causes and previously unrecognized relationships that influence the perceived value of the solution to subpopulations.

Improvement of conceptual artifacts occurs by expanding the facts known about the conceptual artifact. In the vernacular, fact typically means information that is true and accurate; implicit also is the assumption that the fact is objectively defined so that every observer will agree on the both reality of the fact and the meaning of the fact. A more sophisticated view of facts recognizes the role that one's perspective exerts on how one senses and interprets facts. In science, a fact is any idea that can be tested; and some are refuted by tests while others are supported by tests. Those facts refuted by observation are probably inaccurate, and those supported by observation are more likely to be true and accurate.

Richard Feynmann served on the commission that investigated the *Challenger* disaster in 1986, and he observed that much empirical evidence had been ignored and that decisions that led to the disaster had been made for political reasons. Feynman concluded his observations, which appeared in an appendix rather than in the full report, with the statement "For a successful technology, reality must take precedence over public relations, for Nature cannot be fooled" (*Report of the Presidential Commission on the Space Shuttle Challenger Accident*, 1986). Progressive discourse requires the group regard reality as Nature intends, and observation is necessary to expand the factual basis of progressive discourse.

Whereas political and personal whim and individual or social preferences may be acceptable reasons for political decision making, the knowledge building that occurs through progressive discourse requires fact. Political decisions are frequently made for sectarian reasons; one decision is preferred by more powerful participants or one decision is more aligned with *a priori* conclusions or preferences than another. Discourse must proceed in a nonsectarian manner. The proposals to improve the conceptual artifact are all subject to test and the group must accept those facts supported by (and reject those facts refuted by) empirical evidence. In the progressive discourse, planners do have the responsibility to differentiate disparate interpretations that result from different perspectives and those that arise from personal or political whim—a difficult task.

Understanding Data

Data is a term that has been subject to broad interpretation in education in recent decades. Ostensibly, data-driven education is undertaken in an attempt to apply the objective and reliable methods of science to education, thus remove doubt about what educators are doing. It has been established that 21st century educators will face demands that their predecessors did not. Compared to the generations of educators who worked in the 20th century, educators who work in the 21st century will require deeper understanding of human learning, more (and more rapidly changing) information that is potential curriculum, rapidly evolving technology tools, and students who change along with that information and those tools. In addition, progressive discourse will require educators apply the skills of educational researchers as they improve their teaching. It will be necessary for educators to ask clear questions, design methods to gather necessary data, and collect and analyze those data to draw reasonable conclusions in a manner that was unusual for 20th century educators.

It has also been established that data is the least sophisticated of the data-information-knowledge-wisdom hierarchy. To be useful in progressive discourse facts must be established beyond the level of data and transform data into information, knowledge, and wisdom; educators must adopt the same relationship with data as educational researchers.

Research in Schools

Scholars of research differentiate academic research from research conducted in local settings. In the 2007 edition of an education research textbook that had been through seven editions since 1963, Gall, Gall, and Borg (2007) separated two chapters (one focusing on evaluation research and one focusing on action research) into a part of the book addressing the applications of research methods to school practice. Specifically, Gall, Gall and Borg indicated "The primary goal of evaluative research is to make judgments of the worth, value, and utility [of] specific programs. . ." (557). Action research is described as "investigations by practitioners to improve their own effectiveness, that of the organization, or education as a whole" (557). Although action research and evaluation research applies to individuals and small localized populations, so it cannot be generalized and is of limited usefulness in other situations, those researchers have the same responsibilities as academic researchers to be ethical as they gather only the necessary data and analyze and report reasonable findings with clarity.

The Joint Committee on Standards for Educational Evaluation is a professional organization that has—for several decades—published guidelines for those creating systems for evaluating various aspects of educational systems. While several sets of standards from the Joint Committee are relevant to 21st century educators, *The Program Evaluation Standards* (Yarbrough, Shulha, Hopson and Caruthers 2011) are most relevant for educators engaged in research to expand fact for progressive discourse. Evaluative research that conforms to those standards will contribute to the factual improvement of curriculum and instruction while gathering and analyzing data in an ethical manner. The program evaluation standards indicate such research should be of relevance to the community and it should be conducted in efficacious manner, so that it provides valid data from which to draw conclusions. Further, educators who engage in it should evaluate the research undertaken in the local community so the process is improved for the future.

Nature of Data

Different problems require different data, which require different methods. All education researchers must understand the nature of the problems they study and the nature of the methods available so that the correct data can be ethically gathered and reliable conclusions can be drawn. Quantitative methods are used to gather numeric data, which are analyzed using statistical procedures to describe populations or to test hypotheses; qualitative methods are designed to gather narrative data, which are interpreted to elucidate hypotheses.

Quantitative data can be used to either describe populations or to establish correlations. Means, ranges, standard deviations, and similar statistics can be used to describe populations and those measures can be communicated using a variety of numeric and graphic representations. Descriptive statistics can be used (for example) to compare performance of different groups on the data from standardized tests that are common in today's schools. Inferential statistics can be used (for example) to establish a correlation between students who experienced an instructional practice and their graduation from high school. Inferential statistics are used to either accept or reject a null hypothesis. Questionnaires and surveys are popular tools for gathering quantitative data and so are checklists or inventories. Scales are useful for quantifying variation. Writing the instrument for quantitative data collection is critical to the success of any research, and researchers must ensure that it both gathers all relevant factors and is written in a language appropriate for the subjects.

In describing a variation of interviewing as a research method, education and social science scholar Irving Seidman (1998) captured the essential focus of qualitative research methods: "understanding the experience of other people and the meaning they make of that experience" (3). Anshelm Strauss and Juliet Corbin (1990) recognized that education and social science researchers may be unaware of the many factors that affect phenomena being studied and that a researcher who ignores important factors when designing instruments to collect quantitative data may introduce unintended bias into the research. These two observations support the conclusion that qualitative methods are appropriate for studying wicked problems in which researchers seek to document the perspectives and experiences of the participants.

Seidman (1998) further suggested the language used by humans to describe experience is a very important source of data, and that textual data gathered through qualitative research methods can be used to explain social phenomena as well as quantitative data can be used to explain phenomena in the natural sciences. Several methods for gathering qualitative research methods have been used in education, and most include an active role for the researcher in collecting narrative data. Interviews (both of individual and of groups) and observation protocols are common as is document analysis. Qualitative data can be presented either as themes or in case studies. When presenting data as a theme, the researcher reads and rereads the narrative data to identify meaningful trends and connecting ideas. The researcher then explicates and supports these themes with reference to the narrative data. When presenting qualitative data as a profile, the researcher selects important elements of the data and creates a narrative description of the observed or interviewed subject.

Quality of Research

As scientists, educational researchers have a professional responsibility to both conduct research that others will judge to be of high quality and to evaluate the quality of others' research. Only that research that is judged to be of high quality becomes incorporated into the knowledge of the field. Because progressive discourse requires fact to be established, educators and other planners of curriculum and instruction have responsibility to assess the quality of research so that tentative conclusions can be replaced with those more supported by observation as they establish fact.

When assessing the quality of any quantitative education research, professionals consider four characteristics of the research: a) internal validity, b) external validity, c) reliability, and d) objectivity. Various textbooks seek to define these terms, but in general, researchers seek to conduct research to identify a real effect rather than an extraneous effect (internal validity) that can be observed in other populations (external validity) by other researchers (objective) using other tools (reliable).

While qualitative researchers seek to produce high quality research, they assess research with terms different from those used by quantitative researchers. Qualitative researchers seek to completely describe phenomena (credibility rather than internal validity) so that their observations can be made in other populations (transferability rather than external validity). Because their methods emerge during data collection, qualitative researchers describe their methods thoroughly (to establish dependability rather than reliability) and objectivity is replaced with confirmability (Hoepfl 1997).

Regardless of the methods they employ, researchers seek to confirm their observations. Typically, triangulation is sought. If the same conclusion can be reached by three different measures and methods, then the hypothesis is supported. In terms of progressive discourse, if there are at least three threads of evidence supporting a clarification or improvement of the conceptual artifact, then it can become the fact upon which the progressive discourse is built. Just as academic researchers seek multiple sources of data to support conclusions, facts that support decisions in progressive discourse require multiple sources of evidence.

Ethics

A part of all education research is recognizing one's responsibility to proceed in a manner that respects the subjects, the process, and the community. Ethical researchers do not endanger the physical or emotional health of subjects, and they take steps to ensure the privacy of subjects and preserve subjects' right to withdraw without penalty. Also, rewarding subjects for participating poses a threat to data and must be avoided. While those lessons are generally learned as part of each educator's initial preparation to enter the field, there are less obvious aspects of being an ethical researcher that emerge when educators approach education as a wicked problem.

When designing solutions to wicked problems, planners understand that each solution matters because individuals' interaction with the solution will be a permanent part of their experience. For researchers, this suggests that any data collection matters and any data collected from a population will influence data collected later. The influences can be exaggerated when data is collected and then reported to subjects. (I have often wondered if my perception that I was not a strong math student was the result of the D that showed up on my report card in fourth grade; most of the students in my classes when I taught math perceived me to be a good math teacher, the data collected and reported about me as a child notwithstanding.)

In research universities, proposals for research are submitted to an institutional review board before any data collection begins. That board reviews the ethics of the methods and ensures that subjects are not endangered and the research will gather the necessary data; until that group is convinced the research is appropriate and ethical, it cannot proceed. While action researchers are unlikely to have access to such rigorous review, they may be well-advised have knowledgeable outsiders review the methods to ensure it is appropriate and ethical.

Transparent Taming

In chapter three "Education as a Wicked Technology," 21st century education was presented as a wicked problem. Whereas tame problems are definable (cause and effect can be clearly identified), understandable (methods for resolving the problem are known or can be known), and consensual (reasonable people will agree on the need to solve it), wicked problems are none of these. As was reviewed in that chapter, the characteristics of wicked problems follow from the observation that the evaluation of solutions depends on one's perspective on the problem and the solution that was implemented in any situation. One strategy to begin the transition to wicked problem solving is to treat wicked problems as if they are tame, but to do so in a transparent manner: Leaders who adopt this strategy take steps to ensure all involved understand the strategy and will create temporary solutions until the problem can be approached in a wicked way. The realities of resource limits in schools necessitates many wicked problems be approached as if they are tame.

Rationale

For educators, the ability to identify when a wicked problem is being tamed is important for several reasons. Others may be seeking to tame a wicked problem, or they may be seeking to solve a wicked problem as if it were a tame problem. In either case, an educator may be using methods developed by and for those who engineer solutions to tame problems. By recognizing when tame approaches are being applied to wicked problems, an educator can take steps to minimize the adverse affects of the solution such as including non-zero-sum solutions rather than zero-sum solutions. Further, by being aware that wicked problems are being tamed, educators can be prepared to be skeptical of using the results of the solution as evidence for expanding fact in progressive discourse. When adopting these less-than-optimal approaches to solving wicked problems, educators have a responsibility to ensure that the method does not result in a different or unintended problem to be solved; educators adopting these methods must be skeptical of the results or conclusions that arise.

Planners who work on wicked problems recognize that the problems are never solved with finality. Each solution causes new problems or results in deeper understanding so new aspects of the problem. The conclusion that a single problem could be the focus of infinite attention while other problems remain is unavoidable. In reality, planners limit wicked problems, so all problems can receive attention.

Limiting the Problem Definition

When planners reduce a complex wicked problem to a smaller one that is easy to define and bound, the planners are locking down the problem definition. It has been established that wicked problems have many and ill defined causes, but when taming the problem through problem lock-down, the planners attempt to solve one cause that is assumed (sometimes accurately, sometime not) to be the primary cause of the problem. This strategy is often adopted when planners recognize the factors over which they can exert control.

Specifying acceptable criteria and asserting the problem is solved when the criteria are met is another strategy for limiting problem definition. If a school seeks to improve students' scores on a particular standardized test, then the expected level may be defined prior to initiating efforts, and once that level is achieved, then school improvement efforts can focus on a different problem. The potential difficulty with adopting this strategy is that the criteria may not accurately and completely define the construct. For example, we may measure "ensure students are career and college ready" by asserting "passing" scores on tests aligned with the *Common Core State Standards* as the criteria. If performance on those tests is not an accurate predictor of success in career and college, then the wicked problem would not be solved despite the assertion.

Although it has been established that wicked problems (and their solutions) are unique, planners may decide that a problem is sufficiently similar to another that a previous solution (or a solution implemented for another setting) will serve as a model for problem solving activities. Such taming of wicked problems may fail, however, as unforeseen factors or unknown assumptions may make the two situations different. The advantage, however, is that solutions can be implemented and then modified to meet the new situation.

Limiting the Solutions

Every solution to a wicked problem is open to interpretation and different populations at different times will evaluate it differently. As a result any solution can be perpetually revised and refined before it is implemented. When caught in that cycle of revision, time and resources are unavailable to address other problems. In this situation, planners can conclude the current solution is sufficient. This allows planners to avoid the situation in which solving a wicked problem consumes a disproportionate amount of any limiting resource or that failure to implement a partially sufficient solution exacerbates the original problem.

When a planner discovers that factors affecting the problem definition (or solution) emerge from levels within the organization over which the planner can exert no control, then the decision may be made to abandon attempts to solve the problem. In this situation, planners develop strategies for accommodating the influences of the problems on their work or develop methods to work around the difficulties so as to minimize the disruption caused by the problem on his or her performance.

A final strategy for taming wicked problems is to define several acceptable solutions and then decide which is the best for the particular situation. This method of taming a wicked problem is similar to defining a criterion of success prior to beginning. In both methods, measures or options are imposed from outside sources; and those solutions may be actually solving a different problem.

CONCLUSION

Education is a well-established social institution. In modern history, it has served the multiple purposes to prepare young people to participate in the economic, political, and cultural life of the society. For many decades, that society was stable and slowly changing with the skills and knowledge needed by one generation was also needed by their children. In that environment, an educational paradigm built upon assumptions about human nature and technology was dominant and that supported a well-defined collection of widely practiced exemplary practices.

In the 21st century, computers and networks and the information available through those devices and over those networks are replacing print as the dominant information technology. In human history, there is evidence that such periods during which one information technology replaces another that was dominant are marked by turmoil as social institutions adjust to new capabilities and new expectations. The current generation of educators is working at a time when we see evidence of that turmoil in schools. The education paradigm that has served many generations with satisfaction is being replaced.

To remain relevant, educators must recognize new assumptions about humans and human learning including new definitions of what it means to be literate and numerate and otherwise prepared for full participation in the emerging society. These new assumptions will cause educators to adopt new habits of mind and their success will depend on new support systems that expect and encourage autonomy, flexibility, shared responsibility, and other unfamiliar positions.

As I was finishing this book and ready to begin editing it for publication, I had occasion to sit with a group of educators, including one person who was new to the field. Our task was to discuss two cases studies, one of a school that our leaders wanted us to model and another of a school that our leaders wanted us to eschew. When it came my turn to share my thoughts, I ranted for several minutes about how both schools we compared were the same as both sought to improve direct instruction so that test scores would improve. I argued that performance on authentic tasks is more indicative of important skills than tests and students can develop essential skills and knowledge even if I don't give direct instruction. I even indicated that, while collectively students' scores on standardized tests scores may be meaningful, my child's individual scores (which are reported to me) are meaningless. Several like-minded colleagues nodded in agreement, but the one individual new to the profession said, "I don't know what to say. If you are right, then everything I know about education is wrong."

If I could talk to her again, I would explain that what she knows about education is not wrong, she understands education as it was in previous generations. She assumed the students entering her class would be the same students that had entered class with her when she was a child. She assumed she would be preparing students for the world she entered as an adult. She assumed that the experiences she had as a student would prepare her students for their future.

In this book, I have presented evidence that supports a new paradigm for education. I have presented a framework that I anticipate will help educators create classroom in which the teaching and learning more accurately reflects students' future than the classrooms I attended (and that my children attended). I have also presented some strategies whereby educators can take an active role in creating effective classrooms within that new paradigm. Those who do not reinvent curriculum and instruction will be marginalized in the coming years in the same way that scientists who deny nature are marginalized.

Educational researchers Scott Garbiner, Cary Aplin, and Gitanjali Ponnappa-Brenner (2007) contrast engineering instruction for well-defined and measurable outcomes with designing instruction for sociocultural environments. Those who engineer instruction seek plans that lead students to meet goals (alternatively they select prescribed instruction plans that are intended to produce the desired outcome), and if the goal is not achieved (and both the time is available, and the educator is so inclined) the instructional plan can be repeated or revised and repeated until the desired goal is achieved. Sociocultural instruction design, on the other hand, proceeds from the assumption that knowledge is built as the result of the interaction of learners with ideas and procedures in complex and ill-defined settings. Rather than learning as the result of highly controlled settings, sociocultural instruction design posits learning results from authentic situations.

In engineered instruction, individuals are all presumed to need the same information and generally it is delivered from a centralized point (the teacher or perhaps a database) to the students. In that case, technology is designed to deliver information. In sociocultural instruction, technology is designed to facilitate discourse and interaction among the students (and the teacher). Whereas engineered instruction is generally amenable to automated evaluation (which is easily accomplished via digital tools), sociocultural learning includes performance-based demonstrations of learning (which can be created and shared via information and computer technology but must be experienced to be evaluated). In sociocultural learning, the tools are embedded in what and how students learn and in how they demonstrate their learning rather than as tools for supporting delivery of information.

REFERENCES

Ackoff, Russell (1989). "From Data to Wisdom". *Journal of Applied Systems Analysis* 16: 3–9.

Allen, Alsdair, and Pete Warden. 2011. Got an iPhone or 3G iPad? Apple is Recording Your Moves." *O'Reilly Media, Inc*. Last modified April 20, 2011, http://radar.oreilly.com /2011/04/apple-location-tracking.html.

Aronson, Elliot. 2003. *The Social Animal*, 5th ed. New York: Worth Publishers.

Arthur, W. Brian. 2009. *The Nature of Technology: What It Is and How It Evolves.* New York: Free Press.

Asao, Olumuyiwa. 2006. "On the Emergence of New Computer Technologies." *Educational Technology & Society* 9(1): 335–343.

Baker, Al. 2014. "Common Core Curriculum Now Has Critics on the Left." *The New York Times*, February 16, 2015. http://www.nytimes.com/2014/02/17/nyregion/ new-york-early-champion-of-common-core-standards-joins-critics.html.

Barlow, John Perry. 1996. "A Declaration of the Independence of Cyberspace." Accessed November 1, 2014, https: // projects.eff.org/~barlow/Declaration-Final.html.

Benkler, Yochai. 2006. *The Wealth of Networks: How Social Production Transforms Markets and Freedom.* New Haven, CT: Yale University Press.

Bereiter, Carl. (2002). *Education and Mind in the Knowledge Age*. New York: Routledge.

Beynon-Davies, Paul. 2007. "Informatics and the Inca." *International Journal of Information Management* 27(5): 306–318. doi:10.1016/j.ijinfomgt.2007.05.003.

———. 2009a. "The 'Language' of Informatics: The Nature of Information Systems." *International Journal of Information Management*, 29(2): 92–103. doi:10.1016 /j.ijinfomgt.2008.11.002.

———. 2009b. "Neolithic Informatics: The nature of Information." *International Journal of Information Management* 29(1): 3–14. doi:10.1016 /j.ijinfomgt.2008.11.001.

———. 2009c. "Significant Threads: The Nature of Data." *International Journal of Information Management* 29(3): 170–188. doi:10.1016/j.ijinfomgt.2008.12.003.

Blumenfeld, Phyllis, Toni Kempler, and Joseph Krajcik. 2006. "Motivation and Cognitive Engagement in Learning Environments." In *The Cambridge Handbook of Learning Science*, edited by R. Keith Sawyer, 475–488. New York: Cambridge University Press.

Bohn, Roger and James Short. 2009. *How Much Information? 2009 Report on American Consumers.* Global Information Industry Center. Accessed December 15, 2014, http:// hmi.ucsd.edu/pdf/HMI_2009_ConsumerReport_Dec9_2009.pdf.

Borg, Linda. (2014). "Common Core Standards for Education Draw Criticism in Some R.I. Communities." *Providence Journal.* Accessed January 18, 2015. http://www.providencejournal.com/breaking-news/content/20140201-common-core-standards-for-education-draw-criticism-in-some-r.i.-communities.ece.

Brown, John Seely and Paul Duguid. 2000. *The Social Life of Information*. Boston: Harvard Business School Press.

Bruce, Bertram and James Levin. 1997. "Educational Technology: Media for Inquiry Communication, Construction, and Expression." *Journal of Educational Computing Research* 17(1): 79–102.

Bruner, Jerome. 1966. *Toward a Theory of Instruction.* Cambridge, MA: Harvard University Press.

Buchanan, Richard. 1992. "Wicked Problems in Design Thinking." *Design Issues* 8(2): 5–21. doi:10.2307/1511637.

Burton, John, David Moore, and Susan Magliaro. 2004. "Behaviorism and Instructional Technology." In *Handbook of Research on Educational Communications and Technology* 2nd ed., edited by David Jonassen, 3–36. Mahwah, NJ: Lawrence Erlbaum Associates, Publishers.

Bush, Vannevar. 1945. "As We May Think." *Atlantic Monthly* 176(1): 101–108.

Bushnell, Don. 1964. "Computers in Education." In *The Revolution in the Schools,* edited by Ronald Goss and Judith Murphy, 56–72. New York: Harcourt, Brace, and World, Inc.

Cambridge, Barbara. 2008. "Scaffolding for Systematic Change." In *Opening Up Education: The Collective Advancement of Education through Open Technology, Open Content, and Open Knowledge,* edited by Toru Iiyoshi and M.S. Vijay Kumar, 357–73. Cambridge, MA: The MIT Press.

Carmean, Colleen, and Jeremy Haefner. 2002. "Mind over Matter: Transforming Course Management Systems into Effective Learning Environments." *EDUCAUSE Review* 46(6): 26–34.

Carr, Nicholas G. 2013. *The Big Switch: Rewiring the World, from Edison to Google.* New York: W.W. Norton & Co.

Casner-Lotto, Jill, and Mary Wright Brenner. 2006. *Are They Really Ready to Work?: Employers' Perspectives on the Basic Knowledge and Applied Skills of New Entrants to the 21st Century US Workforce.* The Conference Board. Accessed January 2, 2015, http:// www.conference-board.org/pdf_free/BED-06-Workforce.pdf.

Castells, Manuel. 1996. *The Rise of the Network Society.* Oxford: Blackwell Publishers Ltd.

Cazden, Courtney. 1987. *Classroom Discourse: The Language of Teaching and Learning.* Portsmouth, NH: Heinemann.

Chickering, Arthur, and Stephen Ehrmann. 1996. *Implementing the Seven Principles: Technology as Lever.* Accessed January 15, 2015, http://www.iupui.edu/~cletcrse/ncaa/seven.htm.

Christensen, Clayton, Michael Horn, and Curtis Johnson. 2008. *Disrupting Class: How Disruptive Innovation Will Change the Way the World Learns.* New York: McGraw-Hill.

Collins, Allan. 2006. "Cognitive Apprenticeship." In *The Cambridge Handbook of Learning Science,* edited by R. Keith Sawyer, 47–60. New York: Cambridge University Press.

Crichton, Michael. 1983. *Electronic Life : How to Think About Computers.* New York: Alfred A. Knopf.

Cuban, Larry. 1986. *Teachers and Machines: The Classroom Use of Technology since 1920.* New York: Teachers College Press.

Davidson, Cathy, and David Goldberg. 2009. *The Future of Learning Institutions in a Digital Age.* Cambridge, MA: The MIT Press.

———. 2010. *The Future of Thinking: Learning Institutions in a Digital Age.* Cambridge, MA: The MIT Press.

Davis, Fred. 1989. "Perceived Usefulness, Perceived Ease of Use, and User Acceptance of Information Technology." *MIS Quarterly* 13(3): 319–340.

Deken, Joseph. 1981. *The Electronic Cottage.* New York: William Morrow and Company.

Deuze, Mark. 2006. "Participation, Remediation, Bricolage: Considering Principal Components of a Digital Culture." *The Information Society* 22(2): 63–75. doi:10.1080 /01972240600567170.

de Vaujany, François-Xavier. 2008. "Capturing Reflexivity Modes in IS: A Critical Realist Approach." *Information and Organization* 18(1): 51–72. doi: 10.1016 /j.infoandorg.2007.11.001.

Dumont, Hannah. 2010. *Educational Research and Innovation the Nature of Learning Using Research to Inspire Practice.* Washington: Organization for Economic Cooperation & Development.

Dunbar, Robin. 1992. "Neocortex size as a constraint on group size in primates." *Journal of Human Evolution* 22 (6): 469–493. doi:10.1016/0047–2484(92)90081-J.

Eynon, Bret, Laura Gambino, and Judith Török. 2014. "What Difference Can ePortfolio Make? A Field Report from the Connect to Learning Project." *International Journal of ePortfolio* 4(1):95–114.

Feenberg, Andrew. 1999. *Questioning Technology*. New York: Routledge.

Freire, Paulo. 1974. *Education for Critical Consciousness.* New York: Continuum Publishing Company.

Furr, Paula, Ronald Ragsdale, and Steven Horton. 2005. "Technology's Non-neutrality: Past Lessons Can Help Guide Today's Classrooms." *Education and Information Technologies* 10(3): 277–287. doi:10.1007/s10639–005–3009–4.

Gall, Meredith, Joyce Gall and Walter Borg. 2007. *Educational Research: An Introduction,* 8th ed. Boston: Allyn & Bacon.

Gallagher, Shaun. 1992. *Hermeneutics and Education.* Albany, NY: State University of New York Press.

Gallimore, Ronald, and Roland Tharp. 1992. "Teaching Mind in Society: Teaching, Schooling, and Literate Discourse." In *Vygotsky and Education: Instructional Implications and Applications of Sociohistorical Psychology*, edited by Luis Moll, 175–205. New York: Cambridge University Press.

Garbiner, Scott, Cary Aplin, and Gitanjali Ponnappa-Brenner. 2007. "Instructional Design for Sociocultural Learning Environments." *e-Journal of Instructional Science and Technology* 10(1). Accessed February 2, 2015, http: //www.ascilite.org.au/ajet/e-jist/docs/vol10_no1 /papers/full_papers/grabinger.htm.

Gazzaniga, Michael. 2008. *Human: The Science Behind What Makes Us Unique.* New York: Ecco.

Gleick, James. 2011. *The Information: A History, a Theory, a Flood.* New York: Pantheon.

Goody, Jack. 1987. *The Interface Between the Written and the Oral.* New York: Cambridge University Press.

Gould, Stephen Jay. 1998. *Leonardo's Mountain of Clams and the Diet of Worms: Essays on Natural History*. New York: Harmony Books.

Gould, Stephen Jay. 1996. *The Mismeasure of Man.* Revised and expanded. New York: Norton.

Greenhow, Christine. 2008. "Connecting informal and formal learning experiences in the age of participatory media: Commentary on Bull et al. (2008)." *Contemporary Issues in Technology and Teacher Education* 8(2). Accessed October 12, 2014, http://www.citejournal.org/vol8/iss3/editorial/article1.cfm.

Gunawardena, Charlotte, Constance Lowe, and Terry Anderson. 1997. "Analysis of a Global Online Debate and the Development of an Interaction Analysis Model for Examining Social Construction of Knowledge in Conferencing." *Journal of Educational Computing Research* 17(4): 397–431.

Hall, Stephen. 2010. *Wisdom: From Philosophy to Neuroscience.* New York: Alfred A. Knopf.

Hannemyr, Gisle. 2003. "The Internet as Hyperbole." Accessed December 20, 2014, http://hannemyr.com/en/diff.html.

Hede, Toby, and Andy Hede. 2002. "Multimedia Effects on Learning: Design Implications of an Integrated Model." In *Untangling the Web: Establishing Learning Links*, edited by S. McNamara and E. Stacey. Proceedings ASET Conference 2002. Melbourne, July 7–10, 2002. Accessed October 18, 2014, http://www.ascilite.org.au/aset-archives/confs/2002/hede-t.html.

Herrington, Jan, Ron Oliver, and Anthony Reeves. 2007. "Authentic Learning on the Web: Guidelines for Course Design." In *Flexible Learning in an Information Society*, edited by Badrul Kahn, 26–35. Hershey, PA: Idea Group.

Hobart, Michael and, Zachary Schiffman. 1998. *Information Ages: Literacy, Numeracy, and the Computer Revolution*. Baltimore, MD: John Hopkins University Press.

Hoepfl, Marie. 1997. *Choosing Qualitative Research: A Primer for Technology Education Researchers. Journal of Technology Education* 9(1). Accessed September 5, 2014, http://scholar.lib.vt. edu/ejournals/JTE/v9n1/hoepfl.html.

Iiyoshi, Toru, and Cheryl R. Richardson. "Promoting Technology-enabled Knowledge Building and Sharing for Sustainable Open Educational Innovations." In *Opening Up Education: The Collective Advancement of Education through Open Technology, Open Content, and Open Knowledge*, edited by Toru Iiyoshi and M.S. Vijay Kumar, 337–55. Cambridge, MA: The MIT Press.

Ito, Mizuko. 2010. *Hanging Out, Messing Around, and Geeking Out: Kids Living and Learning with New Media*. Cambridge, MA: The MIT Press.

Ivinson, Gabrielle. 2000. "The Development of Children's Social Representations of the Primary School Curriculum." In *Social Interaction in Learning and Instruction: The Meaning of Discourse for the Construction of Knowledge*, edited by Helen Cowie and D. van der Aalvoort, 67–92. Kidlington, Oxford: Elsevier Science.

Jeyaraj, Anand, and Rajiv Sabherwal. 2008. "Adoption of Information Systems Innovations by Individuals: A Study of Process Involving Contextual, Adopter, and Influencer Actions." *Information and Organizations* 18(3): 205–34. doi:10.1016 /j.infoandorg.2008.04.001.

Johannessen, Jon-Arild. 2008. "Organisational Innovation as Part of Knowledge Management." *International Journal of Information Management* 28(5): 403–412. doi:10.1016/j.ijinfomgt.2008.04.007.

Johnson, Steven. 2006. *Everything Bad is Good for You: How Today's Popular Culture is Actually Making us Smarter*. New York: Riverhead Books.

Jonassen, David. 1999. *Learning with Technology: A Constructivist Perspective*. Upper Saddle River, NJ: Merrill.

———. 2000. *Computers as Mindtools for School: Engaging Critical Thinking*, 2nd ed.. Upper Saddle River, NJ: Merrill.

Kelly, Kevin. 2010. *What Technology Wants*. New York: Viking.

Kirschner, Femke, Fred Pas, and Paul Kirschner. 2009. "Individual and Group-Based Learning from Complex Tasks: Effects on Retention and Transfer Efficiency." *Computers in Human Behavior* 25(2): 306–314. doi: 10.1016/j.chb.2008.12.008.

Kozma, Robert, Elain Chin, Joel Russell, and Nancy Marx. (2000). The Role of Representations and Tools in the Chemistry Laboratory and Their Implications for Chemistry Education. *Journal of the Learning Sciences* 9(2), 105–143.

Kraushaar, James, and David Novak. 2010. "Examining the Affects of Student Multitaksing with Laptops During the Lecture." *Journal of Information Systems and Education* 21(2), 241–51.

Kuhn, Thomas. 1970. *The Structure of Scientific Revolutions*, 2nd ed. enlarged. Chicago: University of Chicago Press.

Landauer, Thomas. 1996. *The Trouble with Computers: Usefulness, Usability, and Productivity*. Cambridge, MA: The MIT Press.

Licklider, Joseph. 1960. "Man-Computer Symbiosis." *IRE Transactions on Human Factors in Electronics* HFE-1, 4–11.

Lieberman, Matthew D. 2013. *Social: Why Our Brains Are Wired to Connect*. New York: Crown Publishers.

Lincoln, Yvonna, and Egon Guba. 1985. *Naturalistic Inquiry*. Thousand Oaks, CA: Sage Publications, Inc.

Lohr, Steve. 2001. *Go To: The Story of the Math Majors, Bridge Players, Engineers, Chess Wizards, Maverick Scientists and Iconoclasts--The Programmers Who Created the Software Revolution*. New York: Basic Books.

Marchionini, Gary. 2008. "Human-Information Interaction Research and Development." *Library & Information Science Research* 30: 165–74. di:10.1016/j.lisr.2008.07.001.

Margolis, Nancy. 2004. *LITE: Learning in a Technology-Rich Environment: A Quality Enhancement lan for North Carolina State University*. Raleigh, NC: North Carolina State University. Accessed September 10, 2014, http://litre.ncsu.edu/pdf/litre_qep.pdf.

Mayer, Richard. 2001. *Multimedia Learning*. New York: Cambridge University Press.

———. 2005. "Cognitive Theory of Multimedia Learning." In *The Cambridge handbook of Multimedia Learning*, edited by Richard Mayer, 31–48. New York: Cambridge University Press.

McCluskey, Neal P. 2007. *Feds in the Classroom: How Big Government Corrupts, Cripples, and Compromises American Education*. Lanham, MD: Rowman & Littlefield.

Medina, John. 2009. *Brain Rules: 12 Principles for Surviving and Thriving at Work, Home, and School*. Seattle, WA: Pear Press.

Mehlenbacher, Brad. 2010. *Instruction and Technology: Designs for Everyday Learning*. Cambridge, MA: The MIT Press.

Moll, Luis, ed. 1992. *Vygotsky and Education: Instructional Implications and Applications of Sociohistorical Psychology*. New York: Cambridge University Press.

Nardi, Bonnie. 2005. "Beyond Bandwidth: Dimensions of Connection in Interpersonal Communication." *Computer Supported Cooperative Work* 14(2): 91–130. doi:10.1007/s10606.

Nasir, Na'ilah Suad, and Victoria Hand. 2006. "Exploring Sociocultural Perspectives on Race, Culture, and Learning." *Review of Educational Research* 76(4): 449–475. doi:10.3102/00346543076004449.

National Research Council, Committee on Undergraduate Science Education (1997). *Science Teaching Reconsidered: A Handbook*. Washington, DC: National Academies Press.

National Research Council. 2000. *How People Learn: Brain, Mind, Experience, and School*. Edited by John Bransford. Expanded ed. Washington, DC: National Academy Press.

NETS Project, and Susan Brooks-Young. 2007. *National Educational Technology Standards for Students*, 2nd ed. Washington, DC: International Society for Technology in Education.

Nye, David. 2006. *Technology Matters: Questions to Live With*. Cambridge, MA: MIT Press.

Pacey, Arnold. (1990). *Technology in World Civilization*. Cambridge, MA: The MIT Press.

Palfrey, John, and Urs Gasser. 2008. *Born Digital: Understanding the First Generation of Digital Natives*. New York: Basic Books.

Papert, Seymour. 1994. *The Children's Machine: Rethinking School in the Age of the Computer*. New York: Basic Books.

Pearson, Matthew, and Bridget Somekh. 2006. "Learning Transformation with Technology: A Question of Sociocultural Contexts?" *International Journal of Qualitative Studies in Education* 19(4): 519–539. doi:10.1080/09518390600773353.

Pink, Daniel. 2005. *A Whole New Mind: Why Right-Brainers Will Rule the Future*. New York: Riverhead Books.

Pinker, Steven. 1997. *How the Mind Works*. W. W. Norton & Company.

Pflaum, William. 2004. *The Technology Fix: The Promise and Reality of Computers in Our Schools*. Alexandria, VA: Association for Supervision and Curriculum Development.

Polanyi, Michael. 1966/ 2009. *The Tacit Dimension*. Chicago: University of Chicago Press.

Postman, Neil. 1993. *Technopoly: The Surrender of Culture to Technology*. New York: Vintage Books.

Quan-Haase, Anabel. 2010. "Self-Regulation in Instant Messaging (IM): Failures, Strategies, and Negative Consequences." *International Journal of e-Collaboration* 6(3): 22–42. doi: 10.4018/jec.2010070102.

Ratey, John. 2008. *Spark: The Revolutionary New Science of Exercise and the Brain*. New York: Little, Brown and Company.

Reich, Robert B. 1992. *The Work of Nations: Preparing Ourselves for 21st Century Capitalism*. New York: Vintage Books.

Report of the Presidential Commission on the Space Shuttle Challenger Accident. 1986. National Aeronautics and Space Administration. Accessed November 1, 2011, http://science.ksc.nasa.gov/shuttle/missions/51-l/docs/rogers-commission/table-of-contents.html.

Rideout, Victoria Ulla Foehr, and Donald Roberts. 2010. *Generation M2: Media in the Lives of 8–18 Year-olds*. Washington, DC: Kaiser Family Foundation. Accessed November 19, 2014, http://www.kff.org/entmedia/upload/8010.pdf.

Rittel, Horst, and Melvin Webber. 1973. "Dilemmas in a General Theory of Planning." *Policy Sciences* 4(2): 155–169. doi:10.1007/BF01405730.

Rhodes, Richard. 1999. *Visions of Technology*. New York: Simon & Schuster.

Roberts, Donald, Ulla Foehr, and Victoria Rideout. 2005. *Generation M: Media in the Lives of 8–18 Year-olds*. Washington, DC: Kaiser Family Foundation. Accessed November 19, 2014. http://www.kff.org/entmedia/upload/Generation-M-Media-in-the-Lives-of-8–18-Year-olds-Report.pdf.

Roblyer, Margaret. 2006. *Integrating Educational Technology into Teaching*, 4th ed. Upper Saddle River, NJ: Pearson/Merrill Prentice Hall.

Rogoff, Barbara. 1990. *Apprenticeship in Thinking: Cognitive Development in Social Context*. New York: Oxford University Press.

Rosen, Larry. 2010. *Rewired: Understanding the iGeneration and the Way They Learn*. New York: Palgrave Macmillan.

Rushkoff, Douglas. 1999. *Playing the Future: What We Can Learn from Digital Kids*. New York: Riverhead Books.

Sacks, Peter. 1999. *Standardized Minds: The High Price of America's Testing Culture and What We Can Do to Change It*. Cambridge, MA: Perseus Books.

Sandholtz, Judith, Cathy Ringstaff, and David Dwyer. 1997. *Teaching with Technology: Creating Student-Centered Classrooms*. New York: Teachers College Press.

Sawyer, R. Keith. 2006. "Introduction: The New Science of Learning." In *The Cambridge Handbook of Learning Science,* edited by R. Keith Sawyer, 1–16. New York: Cambridge University Press.

———. (2006b). "The Schools of the Future." In *The Cambridge Handbook of Learning Science,* edited by R. Keith Sawyer, 567–580. New York: Cambridge University Press.

Scarmelia, Marlene, and Carl Bereiter. 2006. "Knowledge Building: Theory, Pedagogy, and Technology." In *The Cambridge Handbook of Learning Science,* edited by R. Keith Sawyer, 97–115. New York: Cambridge University Press.

Schmidt, Stanley. 2008. *The Coming Convergence: Surprising Ways Diverse Technologies Interact to Shape Our World and Change the Future.* Amherst, NY: Prometheus Books.

Schofield, Janet Ward. 1995. *Computers and Classroom Culture*. Cambridge, MA: Cambridge University Press.

Seels, Barbara, Karen Fullerton, Louis Berry, and Laura Horn. 2004. "Research on Learning from Television." In *Handbook of Research on Educational Communications and Technology* 2nd ed., edited by David Jonassen, 249–334. Mahwah, NJ: Lawrence Erlbaum Associates, Publishers.

Seidman, Irving. 1998. *Interviewing as Qualitative Research: A Guide for Researchers in Education and the Social Sciences*, 2nd ed.. New York: Teachers College Press.

Seiter, Ellen. 2005. *The Internet Playground: Children's Access, Entertainment, and Mis-education.* New York: Peter Lang Publishing Group.

Skyttner, Lars. 2005. *General Systems Theory: Problems, Perspectives, Practice,* 2nd ed. Singapore: World Scientific Publishing Company Inc.

Small, Gary, and Gigi Vorgan. 2008. *iBrain: Surviving the Technological Alteration of the Modern Mind.* New York: Collins Living.

Snyder, Thomas, Sally Dillow, and, Charlene Hoffman. 2008. *Digest of Education Statistics: 2007.* Washington, DC: U.S. Department of Education. Accessed November 15, 2011, http:// nces.ed.gov/programs/digest/ d07/index.asp.

Snyder, T.D., and Dillow, S.A. 2013. *Digest of Education Statistics 2012* (NCES 2014–015). Washington, DC: National Center for Education Statistics, Institute of Education Sciences, U.S. Department of Education.

Strauss, Anshelm, and Juliet Corbin. 1990. *Basics of Qualitative Research: Grounded Theory Procedure and Techniques*. Newbury Park, CA: Sage Publication, Inc.

Strauss, Valerie. 2014. "6 Reasons to Reject Common Core K-3 Standards — and 6 Rules to Guide Policy." *The Washington Post,* May 2. http://www.washingtonpost.com/blogs/answer-sheet/wp/2014/05/02/6-reasons-to-reject-common-core-k-3-standards-and-6-axioms-to-guide-policy/.

Tapscott, Don. 2009. *Grown Up Digital: How the 'Net Generation is Changing Our World.* New York: McGraw-Hill.

Taylor, Timothy. 2010. *The Artificial Ape: How Technology Changed the Course of Human Evolution.* New York: Palgrave Macmillan.

Teo, Timothy, and Jan Noyes. 2008. "Development and Validation of a Computer Attitude Measure for Young Students (CAMYS)." *Computers in Human Behavior* 24(6): 2659–67. doi:10.1016/j.chb.2008.03.006.

Thomas, Douglas, and John Seely. Brown. 2011. *A New Culture of Learning: Cultivating the Imagination for a World of Constant Change.* Lexington, KY: CreateSpace Independent Publishing Platform.

Thompson, Derek. 2014. "The Thing Employers Look For When Hiring Recent Graduates." *The Atlantic*. August 19. http://www.theatlantic.com/ business/ archive/2014/08/the-thing-employers-look-for-when-hiring-recent-graduates/378693/.

Toffler, Alvin. 1980. *The Third Wave*. New York: William Morrow and Company, Inc.

Tomasello, Michael. 2014. *A Natural History of Human Thinking*. Cambridge, MA: Harvard University Press.

Tucker, Patrick. 2013. "What Is Personal Data and How Much Personal Data Exists?" MIT Technology Review. May 7. http://www.technologyreview.com/news/ 514351/ has-big-data-made-anonymity-impossible/.

Turkle, Sherry. 1995. *Life on the Screen: Identity in the Age of the Internet*. New York: Simon & Schuster.

Unsworth, John. 2000. "Scholarly Primitives: What Methods do Humanities Researchers Have in Common, and How Might Our Tools Reflect This?" Paper presented at the Symposium Humanities Computing: Formal Methods, Experimental Practice, King's College, London, May 13, 2000. Accessed November 19, 2014, http://www3 .isrl.illinois.edu/ ~unsworth//Kings.5–00/primitives.html.

van Merriënboer, Jeroen, and John Sweller. 2005. "Cognitive Load Theory and Complex Learning: Recent Developments and Future Directions." *Educational Psychology Review* 17(2): 147–177. doi:10.1007/s10648–005–3951–0.

van Merriënboer, Jeroen, Paul Kirschner, and Liesbeth Kester. 2003. "Taking the Load Off a Learner's Mind: Instructional Design for Complex Learning." *Educational Psychologist* 38(1): 5–13.

Venkatesh, Viswanath., Michael Morris, Gordon Davis, and Fred Davis. 2003. "User Acceptance of Information Technology: Toward a Unified View." *MIS Quarterly* 27(3): 425–478.

Vermeij, Geerat. 2010. *The Evolutionary World: How Adaptation Explains Everything from Seashells to Civilization*. New York: Thomas Dunne Books.

Vygotsky, Lev. 1978. *Mind in Society: The Development of Higher Psychological Processes*, 14th ed. Cambridge, MA: Harvard University Press.

Wenger, Etienne. 1999. *Communities of Practice: Learning, Meaning, and Identity*. Cambridge, MA: Cambridge University Press.

Wenger, Etienne, Nancy White, and John Smith. 2009. *Digital Habitats: Stewarding Technology for Communities*. Portland, OR: CPsquare.

Wexler, Bruce. 2008. *Brain and Culture: Neurobiology, Ideology, and Social Change*. Cambridge, MA: The MIT Press.

Williamson, Ben. 2013. *The Future of the Curriculum: School Knowledge in the Digital Age*. The John D. and Catherine T. Macarthur Foundation Reports on Digital Media and Learning. Cambridge, Massachusetts: MIT Press.

Willis, Judy. 2006. *Research-Based Strategies to Ignite Student Learning: Insights from a Neurologist and Classroom Teacher*. Alexandria, VA: Association for Supervision & Curriculum Development.

Workman, Michael. 2008. "An Experimental Assessment of Semantic Apprehension of Graphical Linguistics." *Computers in Human Behavior* 24(6): 2578–96. doi: 10.1016/ j.chb.2008.02.022.

Yarbrough, Donald, Lyn Shulha, Rodney Hopson, and Flora Caruthers. 2011. *The Program Evaluation Standards: A Guide for Evaluators and Evaluation User*, 3rd ed. Thousand Oaks, CA: Sage Publications, Inc.

INDEX